THE SOCIETAL IMPACT OF TECHNOLOGY

The Societal Impact of Technology

SAVVAS A. KATSIKIDES
Department of Social and Political Science
University of Cyprus, Nicosia

Routledge
Taylor & Francis Group

LONDON AND NEW YORK

First published 1998 by Ashgate Publishing

Reissued 2018 by Routledge
2 Park Square, Milton Park, Abingdon, Oxon, OX14 4RN
52 Vanderbilt Avenue, New York, NY 10017

Routledge is an imprint of the Taylor & Francis Group, an informa business

Publisher's Note
The publisher has gone to great lengths to ensure the quality of this reprint but points out that some imperfections in the original copies may be apparent.

Disclaimer
The publisher has made every effort to trace copyright holders and welcomes correspondence from those they have been unable to contact.

A Library of Congress record exists under LC control number: 9872618

ISBN 13: 978-1-138-36693-0 (hbk)
ISBN 13: 978-1-138-36696-1 (pbk)
ISBN 13: 978-0-429-42952-1 (ebk)

Contents

The Author   *vii*

Preface   *ix*

Acknowledgements   *xiii*

1   Sociological Context and Information Technology   1

2   Sociology and the Functions of Technological Autonomy   19

3   Gender and Social Construction of Technology   31

4   Organizational Implications of Technological Change:
     Some Aspects of Applied Systems in Hospitals   37

5   The Making of Global Cities: Technology and Social
     Sustainability   51

6   Technological Change and Employment in Society   59

7   Structuring Technology and Working Environments   73

8   Critical Issues for the Domains of an Information Society 85

9   An Interpretation of Sociology in the Information Society 99

The Author

Savvas Katsikides is Associate Professor of Sociology at the University of Cyprus, Department of Social and Political Science in Nicosia, Cyprus. He studied social sciences at the Johannes Kepler University in Linz, Austria. He taught at the Vienna University of Technology (TU), and was a visiting professor at the Leeds Metropolitan University, UK and at the Central Connecticut State University, New Britain, USA. His areas of interest are the sociology of technology and the sociology of work. His most recent publication is *International Perspectives on Informations Systems*, S.Katsikides and G. Orange, Ashgate, Aldershot.

Preface

The book 'The Societal Impact of Technology'collects a number of articles dealing with sociology and technology. Some of the articles have found their way to different journals as Innovation, The European Journal of Social Sciences, International Review of Sociology, Fachzeitschrift Informatik Forum, etc. The Author believes that in this form as a book will offer contributions which will certainly lead to wider discussions of the above issues. In this respect it is hoped that this collection of studies will hold some interest for those who are inter and transdisciplinarily involved both in social and computer sciences.

Furthermore the contributions are of analytical and critical value concerning vital research issues within the context of the emerging information age. The central idea was to draw together research which is devoted to key questions examining the relationship between the various and widely discussed new developments of technological systems and their social impacts.

The social foundations of technology is a newly established interest area of sociology and of course, contributions have only recently been made into this area. On the other hand, increasing interest and research into information highways and their euphorical assumptions is creating a wide spectrum of societal criticism. Computer supported work for instance has led to the development of innovative organizational processes based on technological developments and new communications paradigms. In particular the focus is on the perspectives of such Networking Entities and their many varied implications on:

In an attempt to define the interplay between sociology and technology is the content of the first article which is entitled 'Sociological Context and Information Technology'. Sociological research related specifically to technology remains faithful to its *functionalist* origins. Theoretical sociology aims to provide solutions to a number of fundamental general problems centered on the concept of social structure, as T. Fararo (1989:9) pointed out. In recent years, much criticism has been addressed to the functional analysis of this particular science. Mulkey's (1976:63) central argument, for instance, was that scientists deviate from some of the putative norms with a frequency that is remarkable if one takes into consideration that these norms are becoming firmly institutionalized. Another criticism has been leveled against methodology and specifically revolves around the question whether terminology has been standardized across all empirical studies. Information technology and its negative or positive impact on society and people is a theme of current debate, as are the implications of information technology at different social levels, namely, international, national and organizational. The aim of this chapter is to look at the various needs and approaches of technology and its place in the scientific tradition. In this respect, we examine what is happening in terms of classical sociology and investigate how solutions can be achieved satisfying different and competing concepts of technologies. Finally, we attempt to redefine the problem of theoretical formulations and describe the ARS (Autonomous Reflex System) model.

The second article is entitled 'Sociology and the Functions of Technological Autonomy'. This provides an overview of sociological work, drawing on recent approaches to technology. It reviews different arguments in regard to the sociology of technology and examines the path to social change. After reviewing those classical and contemporary theories which define the impact of technological processes, this chapter explores the effects of sociological theory on technology and suggests that technological change derives from different societal traditions and as such accumulates and reflects social processes and cultures. It is argued that these views could build the basis for analysis of social conflicts in industrialized countries.

The third article, 'Gender and the Social Construction of Technology', deals with the specific sociological research and interpretations concerning gender and technology. It examines the theoretical implications of formal to applied sociology and, at the same time the growing impact of virtual technology and its influence on gender.

The fourth contribution, 'Organizational Implications of Technological Change: Some Aspects of Applied Systems in Hospitals' attempts to throw light on the relationships between technology and communication and on the other side the emerging anthropocentric approach. The continually changing organizational structures, for instance due to the new technological approach or relocation of production plant, prove that a paradigm change of the enterprises' conception is underway. The evaluation and comparison of these above mentioned systems will be presented here. This article problematizes the notion of computerization systems and their organizations in hospitals. The implementation of computers in the 1970s in hospitals brought about many changes in their organizational structure. Furthermore, it can be said that computerizing nursing care units has become a key step for rationalism and control within the medical profession, which includes also nursing practices. A problem, however, was always the term productivity in hospitals, which would be better off if it could be linked with the whole social function of these institutions.

In the fifth chapter 'Control, Power and Social Sustainability in Cities', the relationship between new information technologies and their innovations regarding tele-activities (such as work, teleports, intelligent buildings, data networks, even virtual mobility) in analyzed in order to resolve problems such as local, regional and national unemployment, education, and transport, within and between growing cities and larger urban areas. The focus is on the following three points: firstly, the analysis of the potential of telecommunication and its technologies to establish and promote environmental, economic and social sustainability in European cities; second, to identify a variety of relevant obstacles hindering sustainability in cities with different social, economic, cultural political and technological backgrounds; third, to understand and manage social and cultural integration/disintegration, particularly in the wake of increasing migration. The paper argues that it is possible to develop a framework which human, social and cultural needs can takes into account thus establishing a social integration perspective.

Chapter six, 'The Shaping of Technological Change in Society', looks at the decision to adopt more advanced technology which consequently has a significant impact on several levels within organizations. The question is whether a company, when adopting new technology, does so to update its production or because there is pressure to innovate. This approach, by contrast, could easily lead to the idea that the market could ultimately gain standardization and affect parallel reactions. The core of this chapter will throw light on the relationship between technology and interorganizational relations with respect to innovations that bring about drastic changes in routine working procedures. While studies and surveys of the implementation of information technology are focused on either positive or negative impacts, this paper will investigate some innovative ideas and approaches offering a template which is based on group dynamics and therefore system eminent.

This chapter, however, after briefly identifying a number of different theoretical approaches to the issue, outlines a framework with in which it is possible to place empirical studies and provides a template for them. It then reports on some of the key empirical studies investigating the relation between the application of new technology and employment. It also provides some useful insights into the debate on the effects of technical change on employment. Furthermore, technology and its control will force people to rethink old models. If this premise is true then it is socially constructed. If technology finally is socially constructed then it can be quit. The passage from shaping to social values via standardization, form the major field of new research for social sciences.

Little work has been done, which explores the ecological and sociological attitudes of engineers and technicians (of both sexes) in large enterprises in Austria. This research note reports on the results of a detailed, quantitative and qualitative study. Chapter seven, 'Structuring Technology and Working Environments' deals with this and concludes by considering the implications of these changes mentioned in the study, and clearly shows that various internal processes, innovations and crises in enterprises have to cope with technological and political decision making for trade unions and their future in Austria.

Chapter eight, which is entitled 'Critical Issues for the Domains of Information Society' draws attention to the need to analyse literature from different disciplines in order to understand different aspects of the same definitions. It focus on different issues of information society, from a critical point of view.

Chapter nine, 'An Interpretation of Sociology in the Information Society' takes a general view of sociological work, drawing on recent approaches on the 'philosophy' of information and communication systems and their theoretical links to social theory. It reviews different arguments about information technology and reexamines contemporary ideas about social theory. This chapter concludes, first, that the post industrial society derives from different scientific and societal traditions and as such accumulates and reflects power and in the end creates a new political and societal condition. Secondly, at the same time, it should be noted that information systems are much more a condition of human action instead of a tradition, which serve as the pathmaker for modernity. Finally, rethinking sociological work as it is, and developing concepts as model simulation are the next steps in sociology, as suggested. Finally, various ideas are tested about interpretation and its role explaining the information society.

Acknowledgements

I should like to acknowledge Professor Mike Campbell in particular for the help, support and many fruitful discussions during my visiting Professorship in his institute, PRI, at the Leeds Metropolitan University (1994-1995). Equally valuable was the advice from Professor Dr Edel Hanappi-Egger from the Technical University in Vienna, and Dr Kyriacos Demetriou from the University of Cyprus all of whose words were invaluable to me.

A special acknowledgement is for Dr Karin Schneider for the contributions regarding the case studies in French hospitals she made to chapter 4, 'Organizational Implications of Technological Change. Some Aspects of Applied Systems in Hospitals'. The article is based on an original paper produced by the author and Dr Schneider for the First Conference on Sociology at the University of Vienna in 1993.

Acknowledgements

1 Sociological Context and Information Technology

Introduction

Sociological research related specifically to technology remains faithful to its *functionalist* origins. Theoretical sociology aims to provide solutions to a number of fundamental general problems centered on the concept of social structure, as T. Fararo[1] (1989:9) pointed out. In recent years, much criticism has been addressed to the functional analysis of this particular science. Mulkey's[2] (1976:63) central argument, for instance, was that scientists deviate from some of the putative norms with a frequency that is remarkable if one takes into consideration that these norms are becoming firmly institutionalized. Another criticism, has been leveled against methodology; and specifically revolving around the question whether terminology has been standardized across all empirical studies. Information technology and its negative or positive impact on society and people is a theme of current debate, as are the implications of information technology at different social levels, namely, international, national and organizational.

The aim of this chapter is to look at the various needs and approaches of technology and its place in the scientific tradition. In this respect, we examine what is happening in terms of classical sociology and investigate how solutions can be achieved satisfying different and competing concepts of technologies. Finally, we attempt to redefine the problem of theoretical formulations and describe the ARS (Autonomous Reflex System) model.

Sociology and Technology

Linking sociology and technology together is both a relatively new and old phenomenon in social sciences. In the 19th century the role of machines had realized more or less a process of automation and therefore modernization for society. In the 20th century, the use of computers and, in particular the use of software is commonly accepted and is widely recognized that it reduces manpower to a common status absent of any social or organizational context

1

which defines them (Salzman 1994:282). Salzman also pointed out that two main fields of distinction are made, while recognizing at the same time differences in power, position, function and organizational structure, which are all central in analyses of the workplace or society. Furthermore, these are dimensions of the social context that are important for analysis of computerization in general and in the design of software in particular. The first point is linked with our analysis in this article, the second approach is analyzed by Salzman and Rosenthal, (1994) in their book *Software by Design*.

In general terms, sociologists examine and interpret scientific and research traditions. At least three such traditions exist within sociology: the conflict theory tradition stemming from Marx and Weber, the Durkheimian tradition with its functionalism and ritual solidarity wings, and the micro-interactionist tradition (Fararo 1989:11). Some sociologists have accepted action theory or structuralism, recognizing the variety of world perspectives and domains of work. The era of sociology and technology has given rise to significant problems, but before discussing them, it is useful to map sociological theory and empirical interpretation.

Social scientific work has been much publicized in the past, and turned into technological discourse by Gehlen (1957), Loewith (1964), Marcuse[3] (1941, 1967), Habermas (1968), Linde (1972, 1982), Ullrich (1979), Bamme et al. (1983), Rammert (1983, 1991, 1993), Hülsmann (1985), Beck (1986), Tschiedel (1989, 1990), Alemann et al. (1993). The issues have principally focused on the development and the implementation of technology, and on the investigation of the further consequences of technology, primarily from a social and ecological perspective. For instance:

(1)   The evaluation of technology or technology assessment and technology monitoring (see Nowotny 1979, Radkau 1983, Halfmann[4] 1984, Jochem 1988).

(2)   The origins of technology and initial analysis of technology through sociological factors which would determine the meaning of certain technological concepts within historically concrete societal development (see Hochgerner and Berka 1994, Rammert 1991, Dierkes and Hoffmann 1992).

(3)   The extension of scientific awareness to socially relevant areas, hitherto unrelated to the implementation and effects of technology upon everyday life, private life, biology and ecology, agriculture, the elderly, health, etc. (Wagner 1991, Weingart 1989, Joerges 1988).

(4)   The constructive approach, and the state as commonly accepted in the USA, the UK and Scandinavian countries. As

Hochgerner and Berka (1994:108) stated, in socio-technical studies of this kind the combined effects of the protagonists, interests, technological and scientific traditions are investigated in the process of making a decision between different 'technical' competing concepts of technology. Leading the way into this field is the work by Bijker et al. (1987) followed for example by Cronberg (1991).

(5)   The shaping of a new era which involves technological and informational reality as a real social awareness.

However, it is difficult to prove that any of the above theses can by itself or in synergy form a common sociology of technology. Hochgerner and Berka (1994) argue that acceptance of interrelating theories in the discipline of sociology of technology will emerge as a natural result of social processes. In the past however, much attention has been paid to the effective implementation of technology as postulated by Adler and Helleloid (1987), and Majchrzack (1986). Accordingly, Salzman (1989) described the process of turning technology into tools, through the examination of the relationship between technology and culture, and by attempting to identify the forms which it may take.

In order to understand the transformations a theory undergoes when it is brought into practical use, it is necessary to analyse the rationality of technology which should preclude all predictable irrationalities. One of the most basic reasons for the division of labour is the ideology of fundamental differences between women and men. The gender question has only recently been taken into consideration and today modern philosophy is dominated by several dichotomies: mind/body, reason/passion and nature/culture, all of which interact with the feminine/masculine dichotomy in complex ways, (Katsikides/Pohl 1994:35)[5]. The associations between women, nature, passion and the body are very influential in contemporary thought and are based (along with theoretical tools) on the different cognitive styles as identified by *Human Computer Interaction* (HCI) which is consequently rebased on these dichotomies. Gender differences have been regarded as pre-determined and natural: for example so that women have restricted choice and are forced to adapt to male values. The alternative is to develop a completely different model of computer or technology usage based on traditional female qualities such as emotion and personal focus.

Clearly every process needs standardisation and control. As technology focused on very strict natural scientific laws it seemed to circumvent the dilemma of exploitation and oppression. Traditional science based on Thomas Kuhn builds on past scientific achievement acknowledged by certain scientific communities, accepting it as the foundation from which to advance. It also accepts as its basis the known classics of science such as Aristotles' Physics, Ptolemy's Almagest, Newton's Principia and Opticks, etc. Before exploring Thomas Kuhn's theory on paradigm, the basic concepts related to the field must first, be defined. Wittgenstein, for example, termed the various parts of human activities as 'games' and we can find similarities between these games and the activities of real human families. The understanding of the term game can only be achieved through the construction of a catalogue of typical cases which possess a minimum of common characteristics, and in which the definition of an activity must not figure. On the other hand, the term 'paradigm' as established by T. Kuhn[6], was closely related to normal science. In his acceptance of the term, he was suggesting that some accepted examples of actual scientific practice - examples which include law, theory application, and instrumentation-provide models for traditions of scientific research. A paradigms acquisition of key characteristics makes more substantive research possible and signals a mature level of development in its particular scientific field (Kuhn 1970:11). For example, with an established paradigm we can relate the growth and social function of technical workers whose work involves the application of science and technology to the development of Taylorism, in addition to the complexity of the new division of labour, and the state in relation to other states, or within a supra-national state (Smith 1990:452)[7].

Furthermore, the social consequences of such complexities can only be understood through examination of the way in which organizations and their structure are changing. Obviously the role of technology has been central, and has ultimately brought about the rational activities of the economic and social order. Max Weber in *Economy and Society*, structured his analysis around the contradictions between the traditional and the non-traditional. He argued that the transition from a predominantly traditional society to a rational society accounted for modernization. Further into the twentieth century however, modern technology and organizations, characterized by sameness and worker alienation, began to be replaced by post-modern/post-industrial technology so that an organization became characterized by its features of diversity and the challenge it could pose to the individual. Marcuse (1941, 1967) was quite accurate in his assertion that technology was a historical, societal project which could not be reduced to a material dimension because the understanding of technology as a social process has slowly become integrated into sociology, despite some opposition (Ogburn 1922, Habermas 1968). Moreover, it is also important to note that if technology is to be seen as an element of social action, or as a process whereby social relationships are to be constituted, then it must be expressed by a theory of social change (Weingart 1989:11)[8]. Thus it becomes obvious that any distinction between models of technology as causal factors of social change and models of social determination of technology must be eliminated. A theory of social action should integrate and comprise both aspects, i.e., societal change and development of technology, as two conflicting processes. With respect to technology and its development, we are in a position to define two possible hypotheses. The first hypothesis relates to the application and dimension of technology (as a result of the development of the natural sciences), wherein societal consequences can be seen as a result of their use or implementation. The representative the first position was Heinz Hülsmann[9] (1985) who stated that technology is possible only under the circumstances of developed societies, for they alone can support and maintain technology. Hülsmann understands the term technology in its relation to structure and situation, both covering the societal formation, technology is a technique *(Technik)* and belongs to natural sciences. Natural sciences are technological by definition and coexist with technology via a specific structural and functional connection. The second hypothesis states that technologies result from processes related to social action that have contributed to the knowledge of natural science.

The transition from the natural sciences to technology implies a simultaneous transition to sociality (Hülsmann 1985:9) i.e., technology is a social reality and it therefore engenders a real sociality. On the contrary, it could be said that technology creates social reality and materializes a real sociality. This happens not afterwards but during social formation. That means, technology forms society.

Effects of New Technology

How does technological development affect organizations? Hoerning[10] (1989) maintains that it is ironic that we do not possess sufficient sociological tools to provide the necessary insight these issues require, since technology has become a decisional factor in societal action. Technology has affected many areas of life, a fact which is generally ignored by the creators of such structures. In this sense however, technological development can be seen as independent action whilst societal cultural action can be regarded as dependent, since they belong to totally different fields of societal reality. Scientific concepts are usually tied to use theories which describe a specific reality. Nevertheless, the validity of these theories is not always proven. H. Willke[11] (1989) argued that if these theories were measured according to their usage alone then we would more or less have second hand theories. It is obvious that once a theory is actually conceived, it becomes easier to measure and evaluate it. Willke (1989:10) argued further that:

during the periods of Marxism and Neo-Marxism, the relationship between capital and work was the main problem; as far as phenomenology is concerned, it is the problem of daily life. The critical theory was liable to the problem of emancipation, and for the theory of action the problem of social action.

Through this Willke believed that it was important to be aware of which theories were popular, as opposed to those which were less important. Many theories which have survived the test of time were successfully measured by Karl Popper's criteria, which specify that theories should allow the construction, examination and falsification of hypotheses.

Sociological theories should fulfil various functions. As Glaser and Straus[12] (1967) stated:

> the interrelated jobs of theory in sociology are: 1) to enable prediction and explanation of behavior; 2) to be useful in theoretical advance in sociology; 3) to be usable in practical applications-prediction and explanation should be able to give the practitioner understanding and some control of situations; 4) to provide a perspective on behavior-stance to be taken toward data; and 5) to guide and provide a style for research on particular areas of behavior. Thus theory in sociology is a strategy for handling data in research, providing modes of conceptualization for describing and explaining.

Adorno[13] stated for instance, 'the application of theory remained uninfluenced by the examining practice. Furthermore, theory and empiricism cannot enter the same continuum'(1972:83). With respect to the empiricism of technological development processes (Hochgerner[14] 1986:11) and with very few exceptions, sociology regards the equipment and facts of technology almost exclusively as societal, external factors. The systematic linking of technical aspects to social factors, together with the consideration of technology as a societal, indigenously produced element implies a transformation of the structures and modes of operation of social relationships on a long-term basis. If these points are borne in mind, should the issue, the tasks, the theoretical and methodological points of sociology be extended and partly revived on this foundation? Moreover, what is happening with complex positions to which a medium-range theory cannot offer satisfactory solutions? At this point, a synthetic theory is required which can neither be postulated in a systemic theoretical form, nor be gained through the state of social theories. The call for a system theory (Miller 1978) has become stronger than ever.

Salzman[15] (1994:282), for instance, has observed that understanding computer technology as it is designed for and used in the workplace requires analysis of organizations, the labour process and technology in general. While some outstanding work has been done on how technology and organizations interact during implementation, there is still a need for further development of a dynamic model of the social process of the creation and use of technology. It therefore makes sense to combine strongly theory, workplace strategies and practical uses.

The new technology and its reproduction and adaptation to the production process affect various sectors of society. One major effect is the so-called societal deregulation of labour relations, and at this point various interests influence the sphere of an employer's activities: the industrial process was integrated within the bourgeois society, which formed the socio-economic system; political developments were established due to this process, and the technical, systemic structure of work was not planned at the beginning of the technological age nor in the previous structures of work. Mumford[16] (1977:493) argued that:

> trade and agriculture, although they had worked slowly, had possessed a freedom and flexibility like no other system before mechanization, because they were able to work manually as opposed to other sectors which were dependent on expensive machines etc. Tools were personal property and selected based on the needs of the respective worker and often were reconstructed, or were custom-made. In contradiction to complex machines these were cheap, replaceable and easily transportable, but, without manpower, effectively worthless.

Later, the creators of the system reduced manpower and its work to secondary status. Certainly, it is not easy to determine although most major decisions on negative developments are made by the technological elites. This however, has not been the focus or interest of mainstream social science. Noble (1984) when contributing some of the first pioneering approaches to technology, pointed out that engineering work is oriented toward developing technology that reinforces the existing political and social order (see also Salzman, 1984).

The Principle of Interpretive Sociology

Motivation and proof of its existence can easily be found in the economic-specific fields, as Wobbe[17] (1986:37). argued, saying that:

> the adaptation of new technology resulted mainly in increasing productivity, cost saving and work saving. This is after all the core of the capitalistic-enterprise rationalization. If the production level remains the same in the firm

then this will lead to the reduction in manpower.

This leads to two possible management concepts in relation to manpower. The first attitude is completely technologically oriented, and management invests only in this direction, using manpower as an aid to machines. Moreover, management demands strict control and centralization of the decision making authority. The second concept is based on an empirical and non-ideological concept, wherein management identifies the weak structures in technology and invests in human competency accepting local, decentralized decisions. Both concepts are based on criteria of economic efficiency. Thus the establishment of an industrial society delivers the framework for such decisions. Interests which reflect these concepts could be realized in the environment of the firm. Technical systemisation of work in industry at the macro structure level is to be seen as a result of these preconditions. A model in which the perspective of a logical construction of industry concludes with their extrovertial and introvertial form, is able to analyse those more (or less) successful structures. Such structures not only exhibit various dynamic aspects, but are also continually changing.

Industries should examine the elements of macro structures which must be developed in autonomous reflex systems (ARS) (Katsikides[18] 1994:47). These can then quasi develop autonomously, become standardized and finally reflect the basic societal conceptions from which they have emerged, so that they will force employers and employees to operate and react similarly.

As far as the interpretive sociological approach is concerned, it might be useful to start with Parsons who has interpreted classical thinkers like Weber, Durkheim and Pareto, establishing the so called analytical action theory. Ferraro also accepts this term in his work *The meaning of General Theoretical Sociology*, which was published by Cambridge University Press, and its simplicity can be focussed on the phenomena of actions understood from the interpretive perspective. Interpretive sociology was able to be separated into several basic distinctions (Feraro 1989:197).

Conclusions

In the following final section, techno-social structures, defined as ARS and based on the above mentioned interpretations will be analyzed. Technology is the expression of creating a system of contents, which gains standardization and therefore increasing compatibility. To analyse the question of technological dynamics a distinction has to be made in order to introduce the ARS concept. The focus is on the organization's structure. As outlined earlier organizations reflect the result of their internal structure, secondly, their external relations and thirdly, the organizational structure itself.

However the implication is how and by which methods may theories be shifted or transmitted into societal processes? How does technological formation influence organizations? It is without doubt that the technical world it was created by humans with special interests. These human activities give rise to effects which have sealed the working environment of social sciences. Hoerning (1989) maintains that it is ironic in our days that we do not possess enough sociological tools to gain sharp views on these issues. That means that technology has proceeded to a decisional factor of societal action. Technology has caused great number of things which are excluded by technicians from their system of looking at things, despite the fact that they, themselves, are the creators of such structures. However, the argument is why can technological development be seen as an independent action and in contrast the societal cultural action as a dependent issue, since both belong to totally different fields of societal reality?

The Structural Perspectives of the Logo Model

Without the pretension of a complete theory, I would like to present three model perspectives which can be applied to various sizes of organizations. The first perspective of the logo model concerns the organizational structure, which can be found in every enterprise. The second extrovertial perspective shows the sphere of action of the enterprise on the outside world. The third introversial perspective covers the sphere of action in the inner life of the enterprise, where all processes are in operation.

As we have seen the structural change of organizations covers more and more enterprises. Public and private organizations are building their organizational structure on a common criterion. As a functional connection of all management and administrative starting points in the inner life of organizations regardless of their size, joint criteria can be observed. The observation of this coherence will be shown in a shaping logo model where the minimum of a organizational structure will be taken as a basis for new technology as the next logical path of other forms. The results of the various adaptations and applications of information technology, i.e. in production such as CAD, CIM, CAM etc. lead to the argument that a) automation and rationalization effects were the first point and b) the final result was the flexibly oriented production. All these systems require a new method of administration of the . A new theme of includes fields and approaches such as data transmission, telecommunication, innovation of production, rationalization of working operation, of employers etc. and as Keen[19] (1990:298) argued:

The very idea of using telecommunications for business innovation means not automating the status quo but explicitly trying to change entire aspects of the organization or its interactions with its business environment. Most businesses recognize they need to have strategies for telecommunication. They can guess at the organizational outcomes or simply go ahead and later deal with unanticipated consequences. But in any case they have to make the choices.

In the same way, Wagner amongst others suggested in 1988 and again in 1991 finding the whole concept of telecommunication, which can be found both in the airways sector as well as in the public administration, social insurance and in hospitals. The effect of rationalization itself shows a certain change in the organization. Keen (1990:301) stated:

This means that telecommunication already affects the core part of their where they work that constitutes their identity is handled. They have already begun a process of quite radical organizational change. In banking, cash management and fee-based services change the corporate banker's skills and roles. They redefine the selling process and customer relationship. They bring the bank's back office and operations and its marketing units together. They alter the basis for compensation and measure of profitability. This is a simple example of the often implicit hidden organizational changes.

It should now be clear that changes in organizations give the reason for other compatible tasks, which are included in the planning and can later be adopted. Hartmann[20] (1986:180) makes the point that the crisis of the administrative work forms the compulsion of continuing production. The second point concerns the administrative operations which must go faster when necessary. The third point is the assistance of the administrative operation to the flexibility of the enterprise; that means a faster collection and distribution of concrete transformations of information and data. It is not surprising therefore that the installed system which affects new organizational structures, was established as an instrument which operates independently of political and social decisions. That implies that the is not in a position to control the system anymore. The decisional parameters lie outside their action fields.

The Extrovertial Perspective

The previous ideas have focused upon the use of technology and its impacts on organizations. In this section the logo model and its development is analysed. The second perspective after the structural perspectives of the logo model, is to be seen on the external action radius of the enterprise, analyses the relationships of the enterprises with the world outside. Examples are mother and daughter companies, relations with the state, the law, trade unions and other interested organizations and last but not least the customer. The compatibility of an enterprise is obvious through the synergies in the level of employment, on the concepts management and finally sales and marketing area.

The Introvertial Perspective

The third observation should be the action sphere which is to be found in the internal structure of the enterprise. At this point it is necessary to analyse trade unions, content of work, working time, collective bargaining, agreements, security, creation of work, economics, etc.

Interests and Participation

Given these explanations, there is a need to confront the above issues in such a way as to know how these aspects can be changed. Changes in the economy generate new conditions for the inner life of organizations which includes trade unions and other relevant associations. Windolf's[21] (1989:367) point is that:

> a former homogenous working class is disintegrated into multitude professional groups, which exist far away from each other according to their professional, economic and political interests; so that a huge bureaucratic organisation can no longer represent them effectively anymore.

The global interests of the firm target on a liberation of the way how employment operates, namely when the workers start to work, finish work, contents of the transformed working actions etc. Galbraith (1972:225) argued that 'these (aspects) are not anymore to be found under the innovation and modernization effect of the free market economy' and by that Windolf (1989:367) emphasizes that trade unions belong to a specific development level of the industrial system. When this level has been superseded, trade unions lose their original powerful position. On the other hand, should participation be directed onto a new course, where it can act against the deregulation process, as described above? Concrete steps could be the construction of a mechanism that secures the right to be informed as employer by the enterprise and by that allows the unions on a local basis to act effectively. In general, an extension of participation of working organizations, creation of work and technology and protection against unlawful dismissal could be the first steps.

The effects of new technology its reproduction and its adaptation in the production process, affect various sectors of society. One main effect is deregulation of labour relations. At this point various interests on the sphere of employer's activities, governing not only the private but also the public sector.

From the historical point of view, the industrial process was integrated into the *bourgeois society*, which at the same time formed the socioeconomic system. Political developments and the formation of political parties, for instance were established due to this process. On the other hand the permanent reorganization of labour in terms of profit maximization had its real goal in the circulation and accumulation of capital, creating at the same time a system where the perpetuation of the existing societal structures gained priority. The technical systemic structure of work was not planned at the beginning of the technical intention nor in the previous forms of work.

With the above general theoretical background, two final comments have to be made as a conclusion to the sociological approach in analyzing information and technological systems. Firstly, technological changes are influenced by power relations, groups and authority, which in return maintain these structures. T. Kuhn, and D. Nobel, R. Kling and H. Salzman gave sufficient indications about the deterministic role of such contextualization. Secondly, perspective is to be found in the software sector. Salzman (1994:311) whilst observing the role of software and power, pointed out finally that software becomes more a tool for the user than a means of control.

Notes

1   Fararo, J.T. (1989), *The meaning of general sociology*, ASA Rose Monograph Series, Cambridge University Press, New York, 1989, pp. 9-11 (also see Collins R. (1985), *Three Sociological Traditions*, Oxford University Press, New York). The main objective of this book is to define the core problems of general theoretical sociology in the context of setting out and illustrating the logic of a nonlinear dynamical social systems framework. Further, after analyzing the outcomes, Farraro continues to formal treatments of action principles and structural analysis. A variety of traditions are drawn upon to treat theoretical problems of order and integration, as well as to examine searchingly problems of formalization and unification in theoretical sociology.

Fararo sees at least three such inner traditions within sociology: the conflict theory tradition stemming from Marx and Weber, the Durchheimian tradition with its functionalist and ritual solidarity wings, and the microinteractionist tradition. From Fararos point of view these three traditions and others like them (such as action theory for instance) are regarded as parts of a more comprehensive research tradition. To emphasize this point, he call them subtraditions and, following Collins, he named their inner variations wings or branches. Further, he treats systems thinking, structuralism, and action theory as subtraditions, which each of them has various branches.

2 Mulkey, M. (1991), *Sociology of Science,* Open University Press, p.63.

3 Marcuse, H. (1941), *Some Social Implications of Modern Technology.* In: Zeitschrift für Sozialforschung, Jg. 9/1941, Nr. 3/41, pp. 414-439.

4 Halfmann, J. (1986), *Die Entstehung der Mikroelektronik. Zur Produktion technischen Fortschritts,* Frankfurt am Main.

5 Katsikides, S. and M. Pohl, (1994), *Dichotomous Thinking, Women and Technology.* In: A. Adam et al (eds), Women, Work and Computerization (A 57) Elsevier Science B.V. North Holland, IFIP. p35.

6 Kuhn, Th. (1970), *The Structure of Scientific Revolutions,* The University of Chicago Press, p.11.

7 Smith, C. (1990), *How are engineers formed?* in: Work, Employment and Society,Vol. 4 Number 3, Sept. p. 452.

8 Weingart, P. (1989), *Technik als sozialer Prozess,* Suhrkamp. Frankfurt am Main, p.11.

9 Hülsmann H. (1985), *Die Technologische Formation,* Berlin, p. 9.

10 Hoerning K. (1989), *Vom Umgang mit der Dingen,* in: P. Weingart, (Ed.) Technik als sozialer Prozess, Suhrkamp, Frankfurt a. M. p.90.

11 Willkes H. (1989), *Systemtheorie entwickelter Gesellschaften,* Juventa, München, p.10.

12 Glaser, B., A. Straus, (1967), *The Discovery of Grounded Theory.* Aldine, p.3.

13 Adorno, T. (1972), *Soziologie und empirische Forschung. In: Adorno et al. Der Positivismusstreit in der deutschen Soziologie.* Darmstadt und Neuwied, pp. 81-101, p.83.

14 Hochgerner J, (1986), *Arbeit und Technik, Einführung in die Techniksoziologie,* Kohlhammer Verlag, Stuttgart. p. 11.

15 Salzman, H. (1994), *The Social Context of Software Design.* in: S. Katsikides, (Ed.), *Informatics, Organization and Society* Oldenburg Verlag, Wien-München. Salzman (Salzman:282) When introducing to his contribution, writes: understanding computer technology as it is designed for and used in the workplace requires analysis of organizations, the labour process, and technology in general. While some outstanding work has been done on how technology and

organizations interact during implementation, there is still a need for further development of a dynamic model of the social process of technology creation and use. His article outlines some of the important elements for such a model developed from empirical analysis of the development and use of mission critical software systems.

16   Mumford L.(1977), *Mythos der Maschine*, Fischer alternativ Verlag, Frankfurt a.M. p. 493.

17   Wobbe, W. (1977), *Menschen und Chips*. Sovec Verlag, Göttingen. p.37.

18   Katsikides S. (1994), *Interests in the Transformation of Organizations*, in:S. Katsikides, (Ed.), *Informatics, Organization and Society*, Oldenburg Wien-München, p. 47. Further about the dynamics of ARS: 'The dynamics of the ARS reflect the result of the changes within this enterprise, its external relations and finally the organizational structure. Reflexive systems proceed more quickly than other sectors or fields of the economy, and thus are able to activate investors, who quickly realize capital returns and profits. ARS affects various activities in the firm as it becomes a tool for rationalization within a short period of time, and as a consequence of which new compatible structures emerge across all its sectors. Beyond this, ARS demonstrates the dependencies of this system, which lead to coercive detention for investments in new technologies. Investments cause within ARS certain rationalizations not to be found in machines, but in humans. In the dynamic crisis potential, following this starting point, the old systems and its couriers will be eliminated. At the same time however, the already established ARS assists in this process by regulating the market'.

19   Keen, P.G. (1990), *Telecommunications and Organizational Choise*. In:Fulk, J./Steinfield, Ch. Organizations and Communication Technology, Sage Publications, London, pp. 298-301.

20   Hartmann, M. (1986), *Strategien und Resultate der Verwaltungsrationaliserung*. In: Journal für Sozialforschung, Heft 2, p.180.

21   Windolf, P. (1989), *Vom Korporatismus zur Deregulierung*. Journal für Sozialforschung, Heft 4. p. 367.

References

Adler, P and Helleloid, D, (1987), *Effective Implementation of Integrated CAD/CAM*: A model, IEEE Trans. Eng.manag. vol. EM-34, no.2.

Alemann, E, et al. (1992), *Leitbilder sozialverträglicher Technikgestaltung*, Obladen.

Bammè A. et al. (1983), *Maschinen-Menschen, Mensch-Maschinen, Grundrisse einer sozialen Beziehung.* Reinbek.

Beck, U, (1986), *Risikogesellschaft: Auf dem Weg in eine andere Moderne*, Frankfurt am Main.

Bertalanffy, Ludwig. (1978), *General Systems Theory.* 6th revised edition, New York.

Bijker et al. (1987), *The Social Construction of Technological Systems. New Directions in the Sociology of Technology*, Cambridge/ Mass-London.

Collins R. (1985), *Three Sociological Traditions*, Oxford University Press, New York.

Cronberg T. (1991), *Danish experiments-social constructions of technology*, Copenhagen.

Dierkes, M,/Hoffmann, U, (eds.), *New Technology at the Outset. Social Forces in the Shaping of Technological Innovations*, Frankfurt/New York.

Europäische Gemeinschaften.(1980), *Kommission der Europäischen Gemeinschaften*, Dokument-Beilage 3/80, Brüssel.

Flusser, Vilém. (1985), *Ins Universum der technischen Bilder*, Göttingen.

Galbraith, J.K. (1972), *The New Industrial State*, New York.

Glaser, B./Straus, A. (1967), *The Discovery of Grounded Theory*, Aldine.

Joerges, B, (1988), *Technik im Alltag*, Frankfurt am Main.

Gehlen, A, (1986), *Die Seele im technischen Zeitalter. Sozialpsychologische Probleme in der industriellen Gesellschaft*, Reinbek.

Habermas, Jürgen. (1968), *Technik und Wissenschaft als Ideologie*, Frankfurt am Main.

Halfmann, J, (1984), Die Entstehung der Mikroelektronik, Frankfurt am Main.

Hochgerner, J, /Berka, G, (1994), *Social environment of technical progress*, in Katsikides et al. *Patterns of Social and Technological Change in Europe*, Avebury, Aldershot.

Hochgerner, J. (1986), *Arbeit und Technik. Einführung in die Techniksoziologie*, Kohlhammer Verlag. Stuttgart.

Hüllsmann, H. (1985), *Die technologische Formation*, Berlin.

Jochem, E, (1988), *Technikfolgenabschaetzung am Beispiel der Solarenergienutzung*, Frankfurt am Main.

Katsikides, Savvas, (ed.) (1994), *Informatics, Organization and Society.* Oldenburg Verlag, Wien-München.

Linde, H, (1972), *Sachdominanz in Sozialstrukturen*, Tuebingen.

Linde, H, (1982), *Soziale Implikationen technischer Geräte, ihrer Entstehung und Verwendung.* In: Jokisch R. (Hg.) *Techniksoziologie*, Frankfurt am Main.

Loewith, K, (1964),   *Das Verhängnis des Fortschritts,* in: H. Kuhn, Wiedmann, F. (Hg.) *Die Philosophie und die Frage nach dem Fortschritt.* Verhandlungen des Siebten Deutschen Kongresses für Soziologie, München.

Majchrzak, A, et al. (1986), *A quantitative assessment of changes in work activities resulting from computer assisted design,* in: Behavior and Information Technology., vol.5. no.3.

Marcuse, Herbert. (1967), *Der eindimensionale Mensch,* Neuwied-Berlin

Marcuse, H.(1941), *Some Social Implications of Modern Technology,* in: Zeitschrift für Sozialforschung, Jg.9, Nr.3/41.

Miller, James. (1978), *Living Systems,* New York.

Mumford, Lewis. (1977), *Mythos der Maschine.* Fischer-alternativ-Verlag, Frankfurt am Main.

Noble D. (1977), *America by Design.* Oxford University Press.

Nowotny H.(1979), *Atomenergie. Gefahr oder Notwendigkeit.* Frankfurt am Main.

Ogburn, W. (1969), (1922), *Kultur und sozialer Wandel,* Neuwied-Berlin.

Radkau, J,(1983), *Aufstieg und Krise der deutschen Atomwirtschaft 1945-1975,* Reinbek.

Rammert. W, (1983), *Soziale Dynamik der technischen Entwicklung,* Obladen.

Rammert. W, (1991), *Entstehung und Entwicklung der Technik, Stand der Forschung zur Technikgenese in Deutschland.* WZB, Berlin.

Rammert. W, (1993), *Technik aus soziologischer Perspektive.* Obladen.

Salzman, H., (1989), *Computer Aided Design: Limitations in Automating design and Drafting,* in: IEEE Trans. on Eng. Manag., vol.36, no.4.

Salzman, H. & Rosenthal, S., (1994), *Software by Design.* Oxford University Press, New York.

Tschiedel, R, (1989), *Sozialverträgliche Technikentwicklung,* Obladen.

Tschiedel, R, (1990), *Die technische Konstruktion der gesellschaftlichen Wirklichkeit,* München.

Ulrich, O, (1979), *Technik und Herrschaft,* Frankfurt am Main.

Türk, Klaus. (1989), *Entwicklungen in der Organisationsforschung,* Ferdinand Enke Verlag. Stuttgart.

Wagner, Ina. (1988), *Intelligente Organisation,* in: Kolm, P. et al. *Konflikt und Innovation in computerunterstützten Organisationen,* Oldenburg Verlag-Wien-München.

Wagner, Ina. et al. (1991), *Das computerisierte Krankenhaus,* Campus Verlag, Frankfurt am Main.

Wagner I, (1991), Women's Voice. *The case of Nursing Information Systems,* in: AI and Society: Special Issue on Gender and Computers in a Cultural Context.

Walton, Richard. (1987), *Innovating to Compete,* Jossey-Bass, CA.

Weber Max, (1978), *Economy and Society,* Berkeley, University of California Press.

Weingart, Peter. (1989), *Technik als sozialer Prozeß*, Suhrkamp. Frankfurt am Main.

Willke, Helmut. (1989), *Systemtheorie entwickelter Gesellschaften.* Juventa, München.

Wittgenstein, Ludwig, (1958), (1953), *Philosophical Investigations,* 2nd edition, Oxford, Blackwell.

2 Sociology and the Functions of Technological Autonomy

This chapter provides an overview of sociological work, drawing on recent approaches to technology. It reviews different arguments in regard to the sociology of technology and examines the path to social change. After reviewing those classical and contemporary theories which define the impact of technological processes, this chapter explores the effects of sociological theory on technology and suggests that technological change derives from different societal traditions and as such accumulates and reflects social processes and cultures. It is argued that these views could build the basis for analysis of social conflicts in industrialized countries.

Introduction

Further examples of two different perspectives on the social shaping of technology are illustrated, namely, the *macrolevel* and the *microlevel*. At the macrolevel, the theory is based on a dynamic view of the social shaping of technology. A proponent of this approach is Salzman (1994), who argued that technology is socially shaped and is part of a larger network of things and people. He and others accepting this view, used this framework, sometimes referred to as a social construction of technology perspective, to build on the traditional studies of science, technology, and society. Thus a number of studies have examined, for instance, the ways in which technology decisions are shaped by nontechnical factors. Research within the emerging field of the social shaping of technology varies quite dramatically in the approaches used, especially in defining the relevant range of social factors. The second perspective, the microlevel, focuses on the usage of technology and is based on the user's perceptions. The conclusion summarizes both the social and the technological impact, and furthermore, it categorizes the research according to varying schools of thought, and shows a new analytical perspective on the sociology of technology on the emerging sociology of information.

Two main factors are key issues for the problems relating to sociology: the individual and the group. Social groups have been defined as collectives of individuals who interact and form social relationships. From the sociology of small group coming to an understanding of larger social collectives, is an effort which has to be shown in sociology, mainly through data. From another point of view, which focuses more on the theory of technological evolution and is based mainly on economic history and anthropology, Bassalla (1988:25) recognizes the larger changes often associated with inventors as well as smaller changes made over a long time period. His theory is rooted in four broad concepts: diversity, continuity, novelty and selection. Diversity can be explained as the result of technological evolution because artifactitious continuity exists, novelty is an integral part of the made world, and a selection process operates to choose artifacts for replication and addition to the existing stock of made things. The argument here is very much related to the first, traditional separation between sociology and technology; i.e., it seems convenient to split up the world into the 'material' and the 'social'.

More precisely, it can be said that theory on the nature and relationship between technology and society is divided into two classic approaches, the technological and the social determinism.

Theoretical Considerations

Another theoretical approach focussed on political technology and designer technology, which led to actor networks and contingent technology. Elster (1983:9-10) for instance, underlined two main approaches: first, that technical change may be conceived as a rational goal-directed activity and as the best choice among a set of feasible changes; secondly, technical change may be seen as the cumulative addition of small and largely random modifications of the production process. Freeman (1987) has contributed to this argument, proposing a third approach to technical change. While he recognizes a domain of validity to the other arguments and agrees that they arise from rational choice and changes in production, he argues further that new combinations of radical innovations related both to major advances in science and technology and to organizational innovations could provide a third dimension. He (Freeman 1987:5) further states that:

> such new technological systems can offer such great technical and economic advantages to a wide range of industries and services that their adoption becomes a necessity in any economy exposed to competitive economic, social, political and military pressures. Increasingly in this century, the world-wide diffusion of such new techno-economic paradigms has dominated the process of technical change for several decades and powerfully influences economic and social developments even though it does not uniquely determine them.

Although the accumulated innovations should actually influence the technical change, no mention is made of existing concepts such as demand pull or technology push and their effect on the technological descriptions. These views, however, can be broken down into two broad categories. First is the theory of the autonomous development of technology (demand pull), posited by those who claim that it is the market and other economic and social influences which primarily determine the scale, rate and direction, and in some cases, even science itself (Freeman 1987:6). Other scholars, such as Schmookler (1966:204) for instance, demonstrated with statistics and figures on patent inventions that the invention activity lagged behind the highs and lows of investment activity. Based on this connection, he wrote that the main stimulus to invention and innovation came from the changing pattern of demand measured by investment in new capital goods. From this point, he went on to argue that external events (proposed to the invention push), for instance, are primarily responsible for the consistency of investments and play the major role in the demand pull theory.

Later, in 1979, Mowery and Rosenberg pointed out that human needs are almost infinite and often long felt, and cannot explain the emergence of a particular invention at a certain time. They also criticized a series of confusing studies undertaken in the 1960s and 1970s, which attempted to show market demand as the force behind innovation. Eventually, Mowery and Rosenberg came to the conclusion that innovation is the result of the interaction between science and technology push factors. This issue, however, which is based on the autonomous development of technology, has been explained in the past by the author (Katsikides 1994).

Technological Uses of the Autonomous Reflex Systems(ARS)

Reflecting on this problem, Katsikides (1994) on the Autonomous Reflex System (ARS), analyzed techno-social structures. Technology is a system of conditions, which reproduces itself in a quasi-autonomous way (and with increasing compatibility). The system develops itself as an autonomous structure, and reflects symbolism and limited identity within the organizational structures of an enterprise. The ARS dynamics reflect the result of the changes and also its external relations. Reflexive systems proceed more quickly than other sectors or fields of the economy, and thus are able to activate investors, who quickly realize capital returns and profits. ARS affects various activities in the firm because within a short period of time it becomes a tool for rationalization, and as a consequence of which, new compatible structures and standardized norms emerge across all its sectors. Beyond this, ARS demonstrates the dependencies of this system, which lead to coercive detention for investments in new technologies. Investments cause certain rationalizations within ARS, which they cannot be found in machines, but in groupware.

The underlying question, however, is how can the technological development be seen as an independent action while the societal cultural action is regarded as a dependent issue, since they belong to totally different fields of the societal reality? The implication is how and by which methods can theories be shifted or transmitted into societal processes? How does technological development influence organizations? Scientific concepts generally use theories to describe reality, although the quality of these theories always presents a problematic issue. The systematic consideration of technical aspects within social facts, the observation of technology as a societal, endogenously produced element or product, involves the transformation of the structures and modes of operation of social relations on a long term basis. Having this statement in mind, should the issue, the tasks, the theoretical and methodological points of sociology be extended and partly revised.

We move now from the theoretical aspect of technology to its real-world applications. Salzman and Rosenthal (1994) have focused specifically on the design of workplace technology and showed how software design and usage leads to essential tasks of engineering, which in turn involve social values. These social values reflect the economic and political dimensions of organizations and provide the basic assumptions that shape individual perspectives on their world of work. Criticism was addressed at an early stage by Lewis Mumford who focussed on the problem of a technological society and autonomous technology. Jacques Ellul, in *The Technological Society* (1964) warned of the technological dominance of human life with ensuing impoverishment of the human spirit. David Noble (1977) stated that although technology is socially determined, little concrete historical analysis exists which illustrates exactly how. Noble's pioneering work developed a growing interest in, and a body of research on include the social shaping of workplace technology. Other useful works in this direction were studies by Bijker, Hughes and Pinch (1987), and Rammert (1993). As Berka and Hochgerner argued (1994), a completely new orientation is required in order to avoid fixation of social scientific terms upon established technical preconditions. Is not enough to research and investigate the development, extension and consequences of technology by using social scientific methods. As this does not lead to a comprehensive sociology of technology, but rather to a *'commentated'* sociology of technology. Those protagonists involved in the mechanization process must be identified according to their social characteristics and the varying degrees of their engagement.

We have now examined several ideas and proposals which constitute the development of technology within the social context. For example, with the introduction and subsequent abandonment of matchlocks in Japan, Noel Perrin[1] (1989) showed very clearly the distinction between social change and the technological push. From the time of the first industrial revolution until today, only a few sociological works have attempted to explain the phenomenon of technology in its social construction. Moreover Marx (1818-1883), Weber (1864-1920) and Durkheim (1858-1917), regarded as the classical sociologists, were all theorists; their findings were based on historical evidence rather than their own research, and using the comparative method.

From Theory to Application

Adorno stated that 'the application of theory remained uninfluenced by the examining practice. Furthermore theory and empiricism cannot enter the same continuum' (1972). With respect to the empiricism of technological development, Hochgerner (1986), found that it was usual in sociology to take technical equipment and facts into account almost exclusively as societal external factors. Few exceptions existed outside of the dominant lines of the discipline.bearing these points in mind, the question arises of whether the issue, the tasks, the theoretical and methodological points of sociology can be extended and partly revived in this respect. Szèll (1994) focused on the relationship between technology and the environment, finally arguing that the challenge is to redirect the tools of sociological analysis to the understanding of the special ramifications of different social organizations and societies and into equations that throw light on the dual problems of environmental destruction and its control (Szèll 1994:10).

With respect to the relevance of the sociology of technology, Berka and Hochgerner (1994) showed that sociology of technology should not and cannot investigate technology itself, but rather a technologically structured technical society, and that societies, structures and features are what make technology such a powerful force. A further point must also be noted. The above ideas could easily suggest that the influence of sociology of technology was significant for the establishment and maintenance of sociology in general. These ideas may even be seen as a reconstruction of sociological thinking concerning technology.

Furthermore, it must be stated here that the sociology of technology has to establish its parameters within the discipline; that is to say it must move from a peripheral position within the science to hold a more central place within sociological knowledge. However, the process of defining these parameters cannot derive from the traditional - the discussion of theories- 'centre' of science. It must be focused on 'research praxis' whereby these limits are defined and continually relocated. Moreover, it is also important to note that if technology can be seen as an element of social action, or as a process in which social relations are to be constituted, then it must be encompassed within the theory of social change (Weingart 1989:10). Of course the problem of explaining social change is not new, it was a central issue in nineteenth-century sociology. It seems however, that the radical social effects of neo-industrialization and technological development of societies are creating fundamental gaps between existing social systems and the new social evolution of technology. No definitive explanation has as yet been given with respect to the second part of this historical hypothesis, namely the theories of revolution. In a comparative perspective Comte, Spencer and Durkheim have developed different aspects of evolutionary theory. Theories of revolutionary social change derive particularly from Marx, who emphasized the importance of class conflict, political struggle and imperialism as the principal mechanisms of fundamental structural changes (Abercrombie et al. 1994:12).

Following that argument it is obvious that the distinction between models of technology as causal factors of social change and models of social determination of technology must be eliminated.

Concluding Remarks

Rather than trying to understand societal transitions via methodological means, or, as Talcott Parsons advocated, using sociology to study the relationship of an individual's experience to society and history, the starting point for the sociology of technology must be through science. A key point, which was made by Hochgerner (1990:86), refers to the 'formative principles' which will create both a source and a framework. Formative principles as theoretical concept cannot offer a normative, fixed and true picture of societal developments, as this concept was developed and is strongest in industrial countries where the principles of hierarchism, objectivism, and growth are at the heart of the theoretical frameworks (and subsequent empirical analysis). However these signs do not pretend to show the direction that development will take, nor do they describe normative expectations in reference to the historic dynamics of civilization. Rather, they represent a practical implementation and, in large social systems, a specifically related further development of the micro-sociological term 'figuration'.

Included in this term are transformation models in which individuals act according to the situation, 'not only using their intellect, but also with their complete body and their every action in their relationship to each other' (Elias1978: 42). Furthermore, Hochgerner and Berka (1994) have explained why hierarchies, objectivism and growth are termed as formative. Claiming that they are intrinsically valid because they each comprise both material and idealistic elements and organizing social facts (laws, customs and traditions, roles, expectations, etc.) such that the societal development they support is in fact secured by their existence. The security of this continued existence allows for remodeling and changes (throughout the society and even concerning its formative principles). For industrial societies which recognize 'growth' as a constitutive necessity, constant change could even guarantee preservation. Such a societal development can only be maintained continuously if it is able to remodel itself by adopting to constant change in a controlled way. This regulating mechanism which controls human behavior according to the specific needs of a certain given societal development, is termed a formative principle. It organizes existence and change in social behavior over several historical eras without itself being restricted to the respective form of that time. On the other hand, objectivism deals with human behavior when this is standardized, or 'functioning'. Finally, the study of what has come to be termed as the sociology of technology does in fact incorporate elements of sociological methods, since they can illustrate social behavior in a regulated societal system where technology and formal foundations create the context and the perspective.

This chapter has attempted to show that there is a variety of theoretical issues which can be directed to mainstream sociology of technology. One common understanding which derives from the research in the field reveals that most sociological studies on technology use the comparative method, and the remaining apply to the field of technology assessment. We have argued that technology reflects the synergy of power and societal processes, and these must be analyzed under the foci of sociology of science or even of the emerging sociology of information. While sociology of information should address a variety of theoretical perspectives that can be directed towards the social phenomenon of information, they alone do not give sufficient insight into the nature of information either as an object of disciplinary discourse or as an object of nature (Balnaves 1993:108). The approach is that an entirely new concept is required and that there is a vital need for improved analysis with respect to the assessment of technological issues. It can be argued that theoretical considerations have to be linked with practical methodology in order to evaluate technological and societal approaches, because different sets of complexities exist between the cultural and the operational aspects of the functional role of technology. However the issue here is more complex, and the argument can be summarized as follows. The first problem relates to methodology, where it is clear that a global approach, whether theoretical or empirical, reaches its limits very quickly. The second problem is a more general issue that refers to all the social sciences: a common direction to resolve common social phenomena is lacking. Thirdly, it can be argued that a new approach is needed, which would focus on a detailed evaluation and provide a synthesis of all the intervening variables involved in the technological discussion. One example of such an approach is the ARS model (Katsikides 1994, see also Chapter One). Without the pretension of a complete theory, it could be argued here that the ARS three-perspective model which can be applied to various organizational structures, builds a first step in this direction. The first perspective of the model concerns the organizational structure of any enterprise. The second perspective shows the external sphere of action of the enterprise; for example, mother and daughter companies, the relationships to the state, the law, and services. The third perspective covers the internal sphere of action of the organization, for example, trade unions, content of work, working time, collective bargaining, agreements, security, creation of work, corporate identity, enterprises, culture, etc.

Finally, technological developments, like other social, economic, and technical approaches, are not socially neutral, and in the end they deal with different traditions (European, US, Scandinavian, Japanese, etc.). As such they accumulate social processes and reflect them, or, as Thomas Kuhn (1970) put it 'a failure to assimilate fully new conditions and technology will strain the existing structures' of society.

Note

1   Perrin, N. (1989), *Keine Feuerwaffen mehr, Japans Weg zurück zum Schwert, von 1543 bis 1879,* athenaeum, see also the original text 'Noel Perrin. (1989), Giving Up the Gun. Japan's Reversion to the Sword, 1453-1879, David R. Godine, Publishers, Inc. Boston, Massachusetts'. One of the well-known unsuccesful technology transfers was the case of those Europeans (Portuguese) who became the first adventurers to visit Japan in 1543. The Europeans brought with them two matchlocks, muzzle-loading hand guns. The introduction of these novelties within a traditional complex system of values within the Japanese society, created such turmoils that the Japanese desided to abandon them until 1855. While, G. Basalla, in his study *The Evolution of Technology* is arguing that the process to adopt weapons Japan found its place immediately. 'The Japanese were so impressed by these primitive firearms that they purchased them on the spot and set their swordsmiths to work duplicating them. Within a decade, gunsmiths all over Japan were turning out firearms in quantity.

The warring feudal factions in Japan, anxious to obtain weapons superior to their swords and spears, encouraged these developments'. Noel Perrin (1978:78), gives a totaly different view of this historical event. After a time of liberalization by the production of guns, the government' s monopoly was so far stabilized in 1625 that the central government decreased the orders to produce weapons, in 1673 e.g. only manufactured 53 big and 334 small matchlocks were produced. In order to maintain tradional life and power in the Japanese society, the government abolished the use of guns and after 1637 and for the next 200 years no wars were carried out with guns in Japan. The history of Japans shows that technological progress in the 17th, 18th and 19th centuries developed at a lower rate than in the West and in other words, Japan undertook a selective steering of technology, and developed other techniques and technologies in other fields. Finally, when in the first January days of 1855 (Noel Perrin, 1978)14 the USS Vincennes from the US Navy reached Tanegashiuma, south of Japan, they realised immediately that the use of guns in Japan was unknown.

References

Abercrombie N, et al.(1994), *Dictionary of Sociology,* Penguin, London.

Adorno, Theodor. (1972), *Soziologie und empirische Forschung,* in: Adorno et al. *Der Positivismusstreit in der deutschen Soziologie,* Darmstadt und Neuwied.

Alemann, E, et al. (1992), *Leitbilder sozialverträglicher Technikgestaltung,* Opladen.

Balnaves M. (1993), *The Sociology of Information,* ANZJS Vol. 29, No.1, March.

Bammé A. et al. (1983), *Maschinen-Menschen, Mensch-Maschinen,* Grundrisse einer sozialen Beziehung. Reinbek.

Basalla G, (1988), *The Evolution of Technology,* Cambridge University Press.

Bijker, W, et al., (1987), *The Social Construction of Technological Systems,* Cambridge/Mass.-London.

Elias, Norbert, (1978), *Was ist Soziologie?,* München, p.142.

Ellul Jacques, (1964), *The Technological Society,* New York, Vintage Books.

Elster Jon, (1983), *Explaining Technical Change,* Cambridge University Press, Cambridge.

Freeman, C., (1987), *The Case of Technical Determinism,* in: R. Finnegan et al., *Information Technology: Social Issues,* The Open University, Hodder & Stoughton, pp. 5-6.

Gehlen, A, (1986), *Die Seele im technischen Zeitalter.* Sozialpsychologische Probleme in der industriellen Gesellschaft, Reinbek.

Habermas, J. (1968), Technik und Wissenschaft als Ideologie, Frankfurt am Main.

Hochgerner, J, /Berka, G. (1994), *Social environment of technical progress,* in Katsikides et al. Patterns of Social and Technological Change in Europe, Avebury, Aldershot.

Hochgerner, J. (1986), *Arbeit und Technik. Einführung in die Technksoziologie,* Kohlhammer Verlag, Stuttgart.

Hülsmann Heinz, (1985), *Die technologische Formation,* Berlin.

Jahn, D., (1994), *The Role of Organizations in the Establishment of Ecological Consensus in Industrialized Countries,* in: U.S. Vedin and B. Hägerhöll Aniansson, Society and Environment, Kluwer Academic Publishers, p. 215.

Katsikides, S, et al. (1994), *Patterns of Social and Technological Change in Europe,* Avebury, Aldershot.

Katsikides, S. (1994), *Informatics, Organization and Society,* Oldenbourg, Wien-München.

Linde, H, (1972), *Sachdominanz in Sozialstrukturen,* Tübingen.

Linde, H, (1982), *Soziale Implikationen technischer Geräte, ihre Entstehung und Verwendung,* in R. Jokisch (Hg.) Techniksoziologie, Frankfurt am Main.

Loewith, K., (1964), *Das Verhängnis des Fortschritts,* in H.Kuhn, F.Wiedmann (Hg.) *Die Philosophie und die Frage nach dem Fortschritt.* Verhandlungen des Siebten Deutschen Kongresses für Soziologie, München.

Markuse, H,. (1967), *Der eindimensionale Mensch,* Neuwied-Berlin.

Markuse, H.,(1941), *Some Social Implications of Modern Technology,* in Zeitschrift für Sozialforschung, Jg.9, Nr.3/41.

McNeill M., (1989), *Research Methods,* Routledge, 2. edition.

Mowery D. and Rosenberg N., (1979), *The influence of market demand upon innovation,* Research Policy, Vol.8 No.2.

Mumford, Lewis. (1977), *Mythos der Maschine,* Fischer-alternativ-Verlag, Frankfurt am Main, p. 493.

Noble D. (1977), *America by Design.* Oxford University Press.

Ogburn, W., (1969), *Kultur und sozialer Wandel,* Neuwied/Berlin.

Rammert. W., (1983), *Soziale Dynamik der technischen Entwicklung,* Obladen.

Rammert. W., (1991), *Entstehung und Entwicklung der Technik,* Stand der Forschung zur Technikgenese in Deutschland. WZB, Berlin.

Rammert. W., (1993), *Technik aus soziologischer Perspektive.* Obladen.

Salzman, H., (1994), *The Social Context of Software Design,* in: Katsikides, S.

(eds.) (1994), *Informatics, Organization and Society,* Oldenbourg, Wien-München.

Salzman H., Rosenthal, S., (1994), *Software by Design.* Oxford University Press, New York.

Schmookler, J., (1966), *Inventions and Economic Growth,* Cambridge, Mass. Harvard University Press.

Széll G., (1994), *Technology, production, consumption and the environment,* International Social Science Journal, June 1994 Vol. 140, Blackwell, Oxford.

Széll G. (1994). *Sociology: State of the Art II, Political, economic and socio-demographic dimensions,* in: International Social Science Journal, Blackwell, Vol. 140, xlvi, No. 2, June 1994.

Tschiedel, R, (1989), *Sozialverträgliche Technikentwicklung,* Obladen.

Tschiedel, R, (1990), *Die technische Konstruktion der Wirklichkeit,* München.

Ulrich, O., (1979), *Technik und Herrschaft,* Frankfurt am Main.

Weingart, P., (1989), *Technik als sozialer Prozess,* Frankfurt am Main.

Willke, H., (1989), *Systemtheorie entwickelter Gesellschaften,* Noventia.

Wobbe, W., (1986), *Menschen und Chips*, Göttingen, p. 37.

3 Gender and Social Construction of Technology

This article deals with the specific sociological research and interpretations concerning gender and technology. It examines the theoretical implications of formal to applied sociology and, at the same time the growing impact of virtual technology and its influence on gender.

From Formal to Applied Sociology

In the past years, conceptual themes and areas in sociology[1] were primarily associated with the works of Emil Durkheim and Max Weber. Other influences in the same era derived from Ferdinand Tönnies, Georg Simmel, Alfred Viekandt and Leopold von Wiese. Their work was analytical and conceptual in character, and their empirical research was much of a fact and finding nature. All of them remained merely theorists. Their approach was better known as formal sociology. For instance G. Simmel in the beginning of the 20th century, has focused and analyzed social relationships. During his studies has merely examined bias of relationships which differed in substance but displayed the same formal properties. In the same view Max Weber (1922, see Abercrombie et al., 1994:172) outlined the basic forms of human interaction which occur in any society, such as power, namely competition and organization. It seems that the main concerns of sociological research remain the same and as in the past and are linked with the issues of authority, restrictions and deprivations. Power, however, seems to be the distinctive perspective in society. Domination by a group is a possible definition, but power is also exercised by individuals. Weber argued that, when the exercise of power was regarded by people as legitimate, it became authority. One criticism of the Weberian approach is that, by its emphasis on agency and decision making, it fails to recognize that non-decision making may also be an exercise of power. However, he underlined the importance of force and defined the state as an institution which had a monopoly of force. Marxist sociology observed power as a consequence of society's class structure, through the inevitable concentration of the ruling class. Power involves class struggle and not simply conflicts between individuals and the analysis of power cannot be undertaken without some characterization of the mode of production. Talcott Parsons on the other hand defined power as a positive social capacity for achieving communal ends; power is analogous to money in the economy as a generalized capacity to secure common goals within a social system.

These attempts to define power deals only with some aspects of the whole, and to some extent, they do not test other critical concepts and theories such as authority, coercion, conflict theory, elitism, hegemony, leadership pluralism, ruling class, participation and sanction, and their inter-implications.

Outcomes of Technology Usage

Why power and to some extent class relationships are connected with male and female characteristics, has yet to be addressed. A commonly accepted notion in the discussion of gender is that human beings are definitely culturally and socially constructed, although each person is biologically determined. Katsikides and Pohl (1994:35-36) pointed out that existing research assumes that feminine/masculine dichotomy is given and is unchangeable. More precisely, they state:

> One of the most basic reasons for the division of labor is the ideology of fundamental differences between women and men. Modern philosophy is dominated by several dichotomies: mind/body, reason/passion, and nature/culture. These dichotomies interact with the feminine/masculine dichotomy in complex ways. The association between women, nature, passion, and the body are very influential in contemporary thought. There are theories about different cognitive styles in Human Computer Interaction (HCI) which are based on these dichotomies. Most theories however, do not question the feminine/masculine dichotomy. Gender differences seem to be given as natural. Women's choice is restricted to either adapting to male values or to developing a completely different model of computer usage which is based on what has been traditionally assumed to be female: emotion and person-centeredness. This gender differentiation is explained by the cognitive approach (Witkin et al., 1962).

That is to say women are field dependent, submissive, and passive, whereas men are field independent, analytic and self reliant. Katsikides and Pohl, criticized this cognitive approach because this theory is not only biased but also wrong, is primarily based on dichotomous thinking, and therefore restricts female activity to a certain cognitive style, even if this style is supposed to be positive. In our analysis we referred also to the work of Rada

(1991:124), who argued similarly for mixed project groups for large software projects. He classified individuals into three main types: a) task oriented individuals who prefer to focus on task, b) interaction oriented individuals who enjoy the presence of co-workers, c) self oriented workers who are motivated by personal success.

Furthermore, Rada observes that women are interaction-oriented rather than task-oriented, the latter being a typical masculine trait. Both roles are complementary, programming teams should therefore consist of an equal mix of both sexes. Rada certainly has no intention of degrading the achievements of female programmers. Nevertheless, he implies that the traditional division of labor is an effective model of work organization and that women are not as talented programmers as men are. In the same work Katsikides and Pohl (1994:136) showed that:

> The question rather is whether categories are defined in a way that inherently accepts the feminine/masculine dichotomy as given, or if they make the formation of gender differences transparent. The first is a rather static approach, whereas the second shows the dynamic nature of gender differences. Such a dynamic approach implies that the historical development of categories and their permanent construction in interaction processes has to be taken into consideration.

Probably more influential than Radas cognitive argument has been the work of Gatens (1991) who identifies the role of philosophical tradition in this context. He pointed to it's limitations:

> Whereas feminists who adopt the women centered approach argue that women should ignore or avoid philosophical tradition, those who adopt the critique of philosophy approach argue that this tradition must be confronted. As it is tradition that has helped form our conceptions of masculinity and femininity, to affirm the value of femininity or female experience without analyzing the historical construction of this experience is to invite failure and repetition.

An example which reflects many of the determinist views is provided by L. White (1949:336) who stated that 'social systems are functions of technologies and philosophies express technological forces and reflect social systems'. Another step has been undertaken by Morgan et al., (1992) who underlined the fact that there has not been very much research in the area of gender differences in HCI. On the other hand, Sheiderman (1992) states that a clear understanding of cognitive style could be helpful for the design of attractive systems adapted to the needs of specific groups but that the results, as far as women are concerned, are not sufficient to define any basic guidelines. One of the reasons for this might be that potential users who can benefit from the results of HCI are predominantly male.

Katsikides and Pohl (1994:41) referred also to aspects such as Computer Supported Cooperative Work (CSCW), and in line with what has been argued so far, the link to the interaction between women and the computer. Another issue which has been analyzed was the social context of female computer usage.

As already indicated, fundamentally similar configurations of technology and the uncritical adoption of technical systems, create images which must somehow be evaluated through the way they are perceived and interpreted. The relationship between the reality and the images of technology if they are connected with the environment is given a new orientation. Jean Piaget (1950) for instance has indicated that human development is carried out in an interrelation between a human being and a social environment. This interrelation takes place through two complementary ways of adapting the world: assimilation and accommodation. The dialectical relationship between these two processes is the driving force for cognitive development.

Conclusion

One of the main tendencies in current uses of technology, is that they do not mirror social reality. Instead they reflect a standard network of conventions. Many aspects of supporting technologies are simply not meeting the social needs of society. The term socio-technical systems, for instance, describes the way how industry would operate efficiently only if both the social need of employees for satisfying tasks and group working and the technical requirements of production were met simultaneously. Research done by the Tavistock Institute of Human Relations (see Abercrombie 1994:408) suggested that production technology could often be redesigned so as to meet social needs without any loss in technical efficiency. An example for this process is the work of Katsikides and Schneider (1994) on computerization and its adoption in the health sector. Wagner (1991:275-289) pointed out that nursing has been among the types of work which historically have been shaped by women. Three related observations stand out in particular as to what is necessary:

(1)   the partial disembedding of social action from local involvement,
(2)   the creation of new locales for individual or collective activity, across professional boundaries, institutions, cultures, and
(3)   deculturalization - the homogenization of thought worlds.

One possible approach to overcome women's problems with the computer would be to develop software systems more congenial to women, as S. K. Damarin (1993:362-370) pointed out. Although this approach could force women again to assume stereotypical modes of behavior. Katsikides/Pohl (1994:42) finally suggest that the use of ethnomethodology as an approach to analyze female computer usage. Ethnomethodology does not assume gender specific differences as given, but rather assumes that different people use different strategies. There must be some mutual agreement about these strategies to make communication possible. These strategies are variable and depend on the needs of those participants in the communication process. In this context, the concept of gender differences can lose its meaning when strategies change and become more and more similar. Sometimes, tendencies to overcome the feminine/masculine dichotomy appear in an unexpected form. Within the text-based virtual reality environments on the Internet, it is possible to pretend to be the opposite gender. This phenomenon is called gender-swapping. If gender can be changed by a single command, then a strict feminine/masculine dichotomy tends to become obsolete.

Note

1 Sociologists are interested in those aspects of human behavior which are the result of the social context in which we live. They do not concentrate on features which are the result of our physical or biological makeup. Sociology stresses the patterns and the regularities of social life which is, most of the time, orderly and largely predictable. (P. McNeill, 1995) But as McNeill put it on the next question, but what do you actually do, the answer is evidence, which has to be collected from the social world around us, and this requires empirical research to be done. Obvious is also the fact that Marx (1818-1883) Weber (1864-1920) and Durkheim (1858-1917) who can be seen as the classical sociologists, were all theorists, their findings were based on evidence from historians and not on their own research. Their method is known as the comparative method, that means sociological research involves the comparison of cases or variables which are similar in some respects and dissimilar in others (N. Abercrombie et al., 1994). At the same time social surveys were conducted by Charles Booth (1840-1917) using a combination of early survey techniques and other less statistical methods. (McNeill) In the 20th century, the Chicago School and the anthropologists studied the way of life by viewing these societies from the inside (participant

observation). Following the Second World War, Paul Lazarsfeld (1901-1970) gave greater emphasis to the importance of data, being as objective as possible.

References

Abercrombie N. et al., (1994), *Dictionary of Sociology,* Penguin, p.172, p.408, p.329-330, see also Mitchell D.G., (1963), *A Hundred Years of Sociology,* Duckworth, London.

Damarin, S,K, (1993), *Where is Women's Knowledge in the age of Information,* in: Kramarae, C. & Spender, D. (Eds.), *The Knowledge Explosion,* New York, Harvester, Wheatsheaf, pp.362-370.

Gatens, M. (1991), *Feminism and Philosophy,* Polity Press, Cambridge.

Katsikides, S, Schneider, K., (1994), *Organisational implications of technological change. Aspects of dynamic forms in applied systems.* In: Katsikides S. et al., (1994), *Patterns of Social and Technological Change in Europe.* Avebury, Aldershot.

Katsikides, S. Pohl, M. (1994), *Dichotomous Thinking, Women and Technology,* in: A. Adams et al., (eds), *Women, Work and Computerization* (A/57), Elsevier Science B.V.(North-Holland) IFIP, pp. 35-36, p.42, p.41, p.136.

Morgan, K., Morris, R., MacLeod, H.,Gibbs, S. (1992), *Gender Differences and Gognitive Style in Human Computer Interaction* in:International HCI Conference 1992, (Proceedings), St. Petersburg, Russia.

Peaget Jean, (1950), *The Psychology of Intelligence*, Routledge and Kegan Paul, London.

Rada, R.(1991), *Hypertext, From Text to Experttext,* London, New York, McGraw-Hill, p.124.

Shneiderman. B. (1992), *Designing the User Interface*, Reading, Mass., Addison-Wesley.

Wagner I, (1991), *Transparenz oder Ambiguität,* Kulturspezifische Formen der Aneignung von Informationstechniken im Krankenhaus. Zeitschrift für Soziologie, 4 pp. 275-289.

White L.(1949), *The Science of Culture*, New York, Farrar, Straus and Giroux, p.336.

Witkin, H.A., Dyk, R.B., Faterson, H.F., Goodenough, D.G., Karp, S.A, (1962), *Psychological Differentiation*, New York, Wiley.

4 Organizational Implications of Technological Change: Some Aspects of Applied Systems in Hospitals

This session problematizes the notion of computerization systems and their organizations in hospitals. The implementation of computers in the 1970s in hospitals brought about many changes in their organizational structure. Furthermore, it can be said that computerizing nursing care units has become a key step for rationalism and control within the medical profession, which includes also nursing practices. A problem, however, was always the term productivity in hospitals, which would be better off if it could be linked with the whole social function of these institutions. Social function and social relations and the role of information technology is an interesting approach, and as M.L.Campbell in an international colloquium in the Austrian Academy of Science (1990:24) pointed out:

> Systematization of nursing knowledge and nursing practice has a history that predates computerization of Canadian hospitals. Computers have sped up the process and extended it for new purposes. Nurses thought that identifying and codifying the knowledge base of their practice would help free them from medical domination, from being seen as simply an extension of medical practice. In profession-enhancing research over the past several decades, nurse scholars, mainly American, have attempted to objectify and capture nurses' knowledge and reformulate it into conceptual models of nursing.

In her interesting contribution, Campbell (1990:34) concludes speaking about the Canadian nursing case, that on the one hand, the new managerial professionalism requires them to speak from the standpoint of cost-efficiency, using systematic methods and procedures that are increasingly structured into their jobs. On the other hand, they are continuously confronted by what she called 'the standpoint of care' represented by nurses who approach their work from a position of clinical expertise. Difficulties arise for these nurse managers as the demands of a clinical approach are increasingly at odds with cost-efficiency requirements. This contradiction goes to the heart of the definition of professionalism in nursing.

The core of this chapter is to attempt to throw light on the relationships between technology and communication and on the other side the emerging anthropocentric approach. The continually changing organizational structures, for instance due to the new technological approach or relocation of production plant, prove that a paradigm change of the enterprises' conception is underway. The evaluation and comparison of these above mentioned systems will be presented here.

Organizational implications were always a characteristic of adaptation of technology. The advances in technology and organization recreate the manufacturing and management system and excluded the non compatible structures by human manpower and organization. To question concepts on how organizations are functioning now, becomes not only a scholastic term for an effective management but for the first time shows serious implications on the practical daily use of computer supported technology. Implementation of technology, requires more skill workers and judgment, initiates new job qualifications which are based on technological innovations, eliminating, on the other hand, the non up-dated compatible human power.

The aim is to analyze the inter-organizational relations and aspects of technology adaptation processes with respect to innovations that bring about drastic changes in the routine working procedures, and in particular in non-profit organizations.

Definitions and Framework

The rapid technological changes and their applications in many sectors of the economy, demands and creates qualifications which can be extended over the life cycles of hardware and software. Much more, qualifications which emerging among the administration of processes are system eminent. How can new knowledge both be adopted and be skilled. What is the current state of these qualifications which are shorter than the life cycles of hard and software. What can be the common point between profit oriented organizations and those which are offering social services as part of the welfare state, such as hospitals? etc. The classical argument that the organization structures of enterprises could be seen as constant, can no longer be seen as acceptable, since other parts of the firms life more and more incorporate external parameters. W. Flusser argued (1985) how the technical pictures emerged, in his work *Philosophy of Photography*, he tries to link this with symbolism which has been shaped in different epochs.

The present study on the social environment of enterprises is meant to be a major target and if we connect this with the working environment, then we can detect three main phases, namely:

(1)   The embedding of manpower into the labor process, in which labor can be seen as a constant and the working environment as a variable. The variables were the tools and the executive manpower.

(2)   The second phase has been developed and shaped, based on the dynamics of the first process and converted the human to a variable. The human was the executive power, of course dependent on the constants of the enterprises' organizational structures. The systematic Taylorism, as well as other management and organization theories, had standardized the inner life of the enterprise.

(3)   The third phase which is supported by a maximum technological equipment, takes with itself humans and tools in the position of variables. It left the continuously changing organization, the macro system, to look like a constant. According to that, the organizational structures of an enterprise follow a dynamic model, in which permanent compatibility allows the increasing mechanization to set foot in the firm.

Giving an example, from the highly-computerized office forced the various activities towards a certain standardization. In 1986, Wobbe emphasized this by saying that:

> EDP is a way of organization or of organizing technology. At the same time a political discussion emerged due to the creation of employment generated by that, because during the process of mechanization and in particular by organizational changes, different groups of workers and working interests will be automatically affected.

This part will deal very briefly with the technical approach of information systems such as Computer Supported Cooperative Work (CSCW), and on the other hand it will empirically demonstrate the practical uses of general information systems in different hospitals.

The transformation of the enterprise strategy is successful within the human component. Consequently the human function is responsible for success and establishes one of the important sources of enterprise strategy. An organization is compatible by these means or successful, when is not static or if we go back to Türk (1989:25) organizations are only rudimentary on a long term stable, much more there are permanently in motion; they never reach their surface stability, neither by balance nor static, but only by movement. New organizational images produce a constantly changing social reality because their duties, due to the changing market variables, must be redefined every time. On the other hand, these processes head for the working class with determined effect.

Much attention has been paid in the past years to the so called effective implementation of technology or the meaning of local knowledge systems in high-tech organizations. On the other hand approaches like skill based design which leads to organizational effectiveness are gaining more and more interest between the aspect how technologies can be turn into tools or symbolism on organizations as an emerging asset of the creation of realities within the firm, based upon their own professional histories, personalities, value systems (Astley 1985) etc. Furthermore, the aspect of the social construction of technological systems among technologists arises and much more people know now that culture, language and ideology are definitely terms of management. Uncertainty of technological systems affected the same in organizational concepts, because organizational structures should become more organic. Flexibility on the other hand, replace efficiency as one of the key objectives of both the manufacturing and service sectors.

Organizational structures tend to be based mainly on computer supported cooperative work (CSCW), and the production idea is also based on sub-products, standard products with high retail and service cost. (i.e., large computer firms). Since different innovative products have found their way to the market, large enterprises react with the same technical and strategic policy and biased to rudimentary organization models. The uncertainty of market it maintains itself and a non static qualification wave from the large enterprises moves down to the small and medium size enterprises. Other and forthcoming aspects on a computer directed economy argue that the buyer is also buying a piece of culture in his enterprise. On that point of view, small firms are more dependent on high paid manpower with advanced qualifications who may have gained this advances and further knowledge in large enterprises. Wagner (Wagner 1990:52) analyzing the cultural differences in appropriating technology under the main title The Unformatted Hospital, positioned three main step analysis which examine how do cultural patterns of coding reality, of cooperating, dealing with uncertainty and conflict, and handling professional boundaries which affect the adoption of computer technology. She used the following three step analysis for her contribution to the International Colloquium at the AAS (Austrian Academy of Science and CNRS) in 1990.

The first step deals with the concept of an organization's culture and, more specifically, hospital cultures. The second part examines computer technology's potential to transform the organization of a hospital. The third section looks at cultural differences in appropriating technology, re-examining case study material collected in two hospitals in France and in the USA.

As the courier of the new technological system, the hardware sector after seeing the evolution in the software sector and loosing their monopolies, decided to catch up using flexible and user friendly systems in order to be more accessible to the user. SMEs were more flexible and quicker to react than the large manufacturers.

A new realization has dawned with respect to the transferable learning processes within the firm and within the international transfer of technology. Its obvious that international strategic alliances, whether in the form of joint ventures, licensing, franchising, exploration and research consortia and minority share holding have existed for a long time. In the 1970s and in the 1980s a new wave of agreements emerged and current estimates show that international licences and joint ventures alone outnumber fully owned affiliates of multinational enterprises by a ratio of 4:1. Whilst such agreements have by no means been restricted to high technology activities, there is no doubt that the networks of strategic alliances, or 'strategic partnering' as they are sometimes called, have grown faster in this field than in others. Focussing now on some results (Wobbe 1986) the professions with a strong increase in percentage evaluation between 1976-1990 in the USA, in some specific areas like technicians for computer maintenance have shown an increase of 147.6 percent, system analysts increased in these 12 years by 107.8 percent, computer operators showed 87.9 percent and programming specialists a 73.6 percent increase. For (Wobbe 1986) the division which is based on the development of the sectoral employment can be divided into three main areas: the industrial core, the service core and the consumption core. Innovative potentials are to be found further in the meta-industrial system, so that the creation of labor leads to the security of the qualifications with relevance to the society. On the other hand Taylorism as the common classical strategy by the implementation of work in the industry, cannot be justified (Wobbe 1986).

The Health Sector Approach

The health care sector has increasingly received political attention at the international level because the financing of currently valid health insurance systems has reached its limits. The first consequence of this evolution was to find a more 'managerial approach' in leading health care institutions with the aim of reducing costs through higher efficiency and rationalization policies. With the maturation of information technology and its successful introduction in the office environment, the health care organizations have turned their attention to the innovative potential of computers as a competitive weapon. On one hand information technology perfectly meets the enormous information need in patient care. On the other hand it is viewed by hospital managers as the ideal support for documenting and thus controlling the production process in hospital industry which is exemplified in the introduction of patient classification systems (Pouvourville 1985).

The objectives for information systems technology are however more complex than the simple generation of activity reports. An integrated information system is expected:

(1)   to provide a balance between antagonist groups within the organization. Medical and administrative powers strive to intervene actively in the development process to preserve their interests

(2)   to coordinate the relations between the medico technical platforms and the medical services, which are characterized by heavy workloads induced by slow communication means

(3)   to simplify the work organization within the wide variety of different functional areas

(4)   to improve the patient accounting procedures

(5)   to provide the ultimate integrated documentation support in form of the computerized medical record and many more desirable properties of equal importance.

Many hospital institutions have adapted information systems for the needs of the administrative department concerns. The system type fitting best their requirements were mostly centralized architectures with solutions that have been imported from other large organizations forms such as can be found in insurance companies. Until recently, the coordination problems between the multiple participants involved in patient care as well as the diversity of their specific information needs have been the principle obstacles to the computerization of the medical field.

Current system developments mainly are focussed on information retrieval problems related to the patient care. The patient record traditionally used as primary working instrument. It can be defined as a unique, individualized reference document. In any case it plays a central role in the negotiation of computer development between hospital administrators, computer engineers and the actors participating in patient care. The computer system reflects to a large extent the need for documentation and communication in through patient record manipulation tools.

The organizational environment of computer applications favors integrated computer architectures, which emphasize the administrative control of patient movements within the hospital institution. The operational systems are for a large part equipped with modules for:

(1)   the atomization of laboratories
(2)   support for personnel planning
(3)   the maintenance of administrative patient data bases
(4)   scheduling of consultations and examinations
(5)   accounting tasks
(6)   automatic registration/control of medical orders
(7)   patient management.

While the realization of permanent computer archives is technically feasible with the memory capacity of current technologies which have become economically attractive, the choice of system development methodology as a management tool has become a critical success criterion. The mutual dependency between system development teams and user groups for the construction of well designed systems has started an intense discussion on the definition for contents of work and appropriate evaluation methods that will emphasize confidence in the system. However in many cases the disorientation on the one and only right way to develop, and the lack of experience in this new market, contribute to the overall impression of disorganization with respect to the introduction policies.

The Case Studies

The hospital sites analyzed in the survey were selected from the public and private sector. For the most part, they were attached to a university teaching hospital center. All institutions had made or were planning to make investments in computing equipment, that besides handling financial management tasks, could also provide support for the organization of a patient's hospitalization from the perspective of the medico-technical professional or the clinician.

The study focussed on the information system development and acquisition policies applied, affecting the main characteristics of the target system as well as the user involvement. Four hospital institutions are discussed. The first case is a pediatric city clinic which runs with an integrated hospital information system that was developed externally by a major computer manufacturer. The second case is a private institution for the treatment of cancer diseases, equipped with an internally developed system. The third case is the surgical department of a cardiological hospital operating with a highly complex local system combining the time-critical demands of the ICU setting with the general care requirements of pre- and postoperative treatment in the ward. The fourth case is a regional hospital center maintaining a loosely coupled decentralized architecture of local networks.

The Pediatric City Hospital

The publicly managed general city hospital was recently opened, and joined two older hospital institutions. A separate data processing department handles all information system related tasks. All ward within the hospital are connected to a central main frame unit. The laboratories are controlled by a separate system linked with the main unit, thus providing total transparency at the user level. The medical secretarial offices are in addition equipped with personal computers. The major requirement for this integrated computer architecture was the optimization of communication between the functional units. This mainly concerns the transfer of documents, laboratory samples, test results, or medical orders. Especially the wards, the laboratories and the administrative units have been included in the concept. A single software package controls the relations between units and the organization of work within a unit. A patient management module coordinates the patient hospitalization, from the admittance in the patient admittance office to the discharge. The medical orders system dispatches requests and delivers results data accordingly.

User participation in the design and selection procedures of the equipment was practically non-existent, with the exception of the chief executives from the medical domain and the hospital direction which were present in the mixed working groups, gathering experts from the supplier company and the government. The original proposed solution had to be adjusted to the local requirements of the french health system. Even with a product that had been approved on the American market, major amendments had to be carried out. The strong 'costing out' design orientation was considered by the users as a primary obstacle, since it is not part of the French working culture in the health care environment. Besides this fundamental issue, the medical professionals had to familiarize themselves more than any other professional group with the system. The orders module forces them to fully exert their prescription role exclusively using the computer tool. The recognition of this change in the working procedures demanded a considerable time and adaptation effort.

The essential difficulty was the poor flexibility of the data processing department in complying with user requests. As a consequence, all user groups organized themselves into working groups. On a regular basis meetings with the data processing staff are scheduled, to discuss proposals or bring forth problems, experiences in the daily work with the computing system. This feedback brought about a significant change in the quality of the application software in use. Although the conceptual foundation cannot be questioned anymore, the concrete interventions of the hospital staff has changed the practice in the selection, modification, design and implementation tasks in the sense, that the affected users intervene already in the design stage. The introduction of this new technology has also brought up resistances among the other professional groups such as the nurses who directly depend on the physician's prescription role.

The City Cancer Clinic

The privately managed Cancer Center in one of the major French cities and is composed of 26 different medical units and technical departments, most of which are connected to a central computing system. In comparison with the pediatric city hospital the size of the computer expert team is relatively modest (one computer system engineer assisted by two full-time and three part-time system analysts). The introduction of the computing system was motivated by a general discontent with the inefficient practice in the maintenance of paper records.

In this institution the system design and implementation process was controlled by the medical staff. The project leader was the medical director of the institution. A cooperation contract with a major supplier of hard and software equipment was made. The computer manufacturer provided a requirements specification and a basis software environment for the generation of a data base. A low cost solution with minimal architectural disruptions was set down.

The first application for the management of the medical record was then written by the local computer professional (who was hired at that time). The presently used system consists of a main frame unit installed in wards, and offices. The major demand on the part of the clinicians for fast access to patient related information was the main functional requirement. The central application module is therefore the medical record, which is structured into a set of folders, typified by the professional group, the medical discipline and functional areas within the hospital. The design of the computerized record was modeled on the paper record.

Access to the medical records is connected to organizational tasks, such as patient admission, discharge, transfer, as well as tasks related to the generation of medical reports or orders. The user interacts through a set of functions and subfolders defined for his profession. The system is well adapted to the medical and secretarial work environment. Gradually modules have been added for the laboratory assistant, and the administration clerk.

Although the implementation work has partly been delegated to the supplier company, the design and implementation policy was one of formal control exerted by the representatives of the hospital institution. Agents from the technical, the laboratory, the secretarial, and the medical staff were always present in the mixed commissions, negotiating with the supplier. They also met (and still do when required) in the event of major changes to the existing hard and software environment. Reactions of refusal with respect to the technology in user were relatively infrequent, due to the general consensus on sharing information with electronic devices and the precautions taken, to develop application which are exactly tailored to the needs of this specific institution. At times user participation also takes place informally in daily problem situations where the data processing professional are contacted and informed about errors, requests, improvements etc.

The Cardiological Clinic

The cardiological clinic maintains a surgical department with two units managing pre and post operative care in the wards, an OP, two intensive care units and a local laboratory. The surgical units are controlled by a local computing center. Patient treatment spans from the preparation to an intervention, through a 48 hour supervision in the ICU and the post operative case in the ward.

A single computer expert is responsible for all system acquisition, maintenance, design and implementation tasks. The present computing architecture evolved from an isolated monitoring system to a highly complex architecture integrating the collection of real-time data (vital signs of the patient) delivered during the postoperative period in the ICU with the medical and especially the nursing practice in the wards. Every stage of the hospitalization is documented in detail on a single patient record.

On the hospital institution scale the system in this department is a decentralized solution, whereas on the departmental level, it is a centralized system. A hybrid central unit (designed for batch, and real-time multi-user work environments) controls the communication between the wards, the ICU, the OP, and the laboratory. The system was originally purchased to handle real-time processing of data in the ICU. Its logic is therefore centered around the efficient storage and display of short-lived data.

The functionality was later extended to integrate the physician's prescription role, the nursing care process, the actions of the physician and the pharmacy stocks. The nursing documentation system is a particularly developed feature. It is entirely based on the knowledge and experience of the nursing staff of the surgical department. The experienced representatives from the medical and paramedical staff participated in sketching out the basis design. In the realization of the computer application, the computer expert was in turn allowed complete autonomy.

The outcomes of his implementation work were presented as prototypes to the users. They were repetitively overworked until final acceptance, by the staff. From the early design stages on, the affected users were intricately involved in the design decisions. Their contributions with respect to the semantics and the presentation of the record, were used as the primary source for the requirements specification. The system is widely accepted and heavily used. Its success is based of the direct and informal relationship between the computer expert and the hospital staff on one side, and on the integration of multiple perspectives into the well comprehended computerized record.

The Regional Hospital Center

The regional hospital complex is structured into a set of distinct buildings arranged in a campus area. In the computerization process these architectural conditions were prioritized. The computing system in use is a decentralized cluster of local networks connected to a central interface component which handles the exchanges between a main frame unit used by the administration and medical units. Presently two thirds of the departments are computerized. Each unit is equipped with a local network of personal computers with one master station and a printing device. The overall software platform is based on two separate applications. The first is an integrated package treating matters of the hospital administration. The second application is centered around the patient record and is embedded in standardized hard- and software environment. Patient records are stored on a commercially available data base system. Information is assigned to three categories ranging from a collection of publicly available hospitalization summaries, to specialized records owned by units and individual physicians.

From the initial design stages on, the future users have been involved in the development process, which was managed by the local computing department within the hospital institution. They intensively contributed in creating the specialization records concept. Their constant participation is an absolute necessity to maintain a regular system operation.

Especially with respect to the local data organization and archival the units are totally independent from the data processing staff. The local data processing experts only intervene in problem situations. The resulting decentralized organization therefore also brings about an equal distribution of knowledge going beyond the simple manipulation of a computer program. In each unit a local system administrator from the medical or paramedical profession has been specially trained to maintain the network. The users can thus seek help from this internal communication partners.

With a special record structure tailored to the requirements of each medical discipline the basic information needs of the physician has been satisfied. However a major source of discontent and poor work overall work performance with respect to the nursing profession, the improvement of the relations with the laboratory sector has gone unnoticed. With the current system functionality only two professional groups (doctors and medical secretaries) have been integrated. The interests of the vast majority of the hospital staff have unformtunately been ignored.

References

Astley, W.C., (1985), *Administrative Science as socially constructed truth.* Administrative Science Quarterly, 30, 497-513.

Campbell, L.M. (1990), *Systematization of nursing and the promise computers,* in: Proceedings of an International Colloquium, Austrian Academy of Science and CNRS, Vienna, Austria.

Bill Curtis, Herb Krasner, Neil Iscoe, (1988), *A field study of the software design process for large systems,* Communications of the ACM, Vol. 31, No. 11, pp. 1268-1287.

Downey, H. Kirk H. Arthur P. Brief, (1986), *How Cognitive Structures Affect Organizational Design: Implicit Theories of Organizing,* in: Henry P. Sims, Jr., Dennis A. Gioia and Associates: The Thinking Organization: Dynamics of Organizational Cognition, pp. 165-190, Jossey-Bass Publisher.

Gioia, A. D. (1986), *Symbols, Scripts, and Sensemaking: Greating Meaning in the Organizational Experience,* in: Henry P. Sims, Jr., Dennis A. Gioia and Associates: *The Thinking Organization:* Dynamics of Organizational Cognition, pp. 49-74. Jossey-Bass Publishers.

Flusser, V., (1985), *Philosophy of Photography,* European Photography, Göttingen.

Katsikides, S, Schneider, K., (1994), *Organisational implications of technological change. Aspects of dynamic forms in applied systems,* in: Katsikides S. et al., (1994), *Patterns of Social and Technological Change in Europe.* Avebury, Aldershot.

Katsikides, S., Pohl, M. (1994), *Dichotomous Thinking, Women and Technology*, in: A. Adams et al., (eds), *Women, Work and Computerization* (A/57), Elsevier Science B.V.(North- Holland)IFIP, pp.35-36, p.42, p.41, p.136.

Lincoln, S. Y. (1985), *Organizational Theory and Inquiry: The Paradigm Revolution*, Sage Publications.

Linda Klebe Trevino, Robert H. Lengel, Richard L. Daft, (1987), *Media Symbolism, Media Richness, and Media Choice in Organizations*, A Symbolic Interactionist Perspective, in: Communication Research, Vol.114 No. 5, October 1987, pp. 553-574, Sage Publications, Inc.

Pouvourville Gerard, (1985), *Hospital System Management in France and Canada:* National Pluralism and Provincial Centralism, in: Sc.Sci. Med. Vol. 20, No. 2. pp. 153-166, Pergamon Press, UK.

Ronald E. Rice, Frederick Williams, (1987), *Theories Old and New: The Study of New Media*, The New Media: Communication, Research and Technology. Sage Publications, Beverly Hills.

Susan Leigh Star, (1989), *Layered Space, formal representations and Control: the politics of information.* Fundamenta Scientiae, Vol. 10, No. 2, pp. 125-155, Brazil.

Wagner Ina, (1990), *The Informed Hospital, Cultural Differences in Appropriating Technology*, in: Proceedings of an International Colloquium, Austrian Academy of Science and CNRS, Vienna.

Weick E. Karl, (1990), *Technology as Equivoque: Sensemaking in New Technologies.* in: Goodman/L.S. Sproull and Ass: Technology and Organizations. Jossey-Bass, San Francisco, Oxford pp. 1 - 44.

Wobbe, W., (1986), *Menschen und Chips*, Sovec, Göttingen.

Zahrly Jan, (1990), Organization Design, in: L.R. Gomez-Mejia/M. W. Lawless, *Organizational Issues in High Technology Management.* Jai Press Inc. pp 79.

5　The Making of Global Cities: Technology and Social Sustainability

This chapter aims to analyze the relationship between new information technologies and their innovations regarding tele-activities (such as work, teleports, intelligent buildings, data networks, even virtual mobility) in order to resolve problems such as local, regional and national unemployment, education, and transport, within and between growing cities and larger urban areas.

The focus, is on the following three points: First, the analysis of the potential of telecommunication and its technologies to establish and promote environmental, economic and social sustainability in European cities; second, to identify a variety of relevant obstacles hindering sustainability in cities with different social, economic, cultural political and technological backgrounds; third, to understand and manage social and cultural integration/disintegration, particularly in the wake of increasing migration. The paper argues that it is possible to develop a framework which human, social and cultural needs can takes into account thus establishing a social integration perspective.

Introduction

Sociological comment on urbanization, urban way of life, urban social movement, and urban ecology, are some of the crucial aspects which concentrate on the new role of cities, the urban areas and the individual. In general, periods of intensive industrial and economic tertiary activities are associated with consequences for labor, uncontrolled growth of cities and, lastly, with unemployment. Karl Mannheim[1] used the term utopia in order to describe the beliefs of subordinate classes, especially beliefs which emphasized those aspects of society which pointed to the future collapse of the established order. Whereas Karl Mannheim suggested that utopian thought would not be characteristic of the twentieth century, some sociologists claim that modern pessimism over, for example, nuclear warfare represents dystopian thought, a collapse of civilization without a subsequent social reconstruction. As certainly utopia is to believe that all social problems could be solved if they arose from urban development or, as Max Weber proposed, through the comparative study of civilizations. In short, Weber[2] went on to trace the development of the urban community into a secular association whose members enjoyed civic rights and received economic advantages in return for their payment of taxes. Such cities differed in many subtle ways from those in other cultures. He pointed out, that some urban developments were more conducive to the development of capitalism than others.

A characteristic model[3] in the cities of the 19th century was their rapid expansion and their subsequent slowing down, of greater suburban development. The proportional increase in urban populations in the 19th century was largely by migration from the countryside. Urbanization has contradictory consequences for economic growth, since it cheapens the cost of providing services such as health and education while increasing the cost of labor that can no longer supplement its wages by small-scale agricultural production. The movement from the urban areas to the accumulated centers of labor and capital, forced people in the 19th century to flee from their traditional places to new environments.

In addition the traditional metaphor of the 'Global Village' in new urban planning and studies about the future of cities, particularly the concepts of the 'Metropolis' and of the 'Global City', provide analytical instruments and insights concerning increasingly relevant social problems of the cities and regional developments.

Contemporary Societies

The labor market, and its connection to unemployment, poverty and refugees are the classical social exclusion patterns in the cities. The role of cities has come to a turning point and must develop itself to a regulation instrument where its social role should be underlined. Cities in the past, were not in a position to cover and develop social programmes, because in several legislations the central governments possessed the power to incorporate these in their central policy approach. Local government policies have taken over. Several sociopolitical and administrative structures, however, have changed; new ideas are in shaping which are connected with mobilization, and the notion of multi-culturalism or the relationship to ethnic minorities. As Rex J. and Y. Samad (1996)[4] stated:

> these issues have often been discussed in general terms and have produced

considerable disagreement. This disagreement, however, has often resulted from the fact that the different parties in the argument have been referring to different situations, and it was agreed amongst some of them that the argument could best be taken further on the basis of empirical studies for the kinds of associations and institutions in major European cities.

Transnational Urbanization

The breakneck speed of information and communication technology and its diffusion along with the increasing information processing to which all society is subjected, is changing not only the fields of material and intellectual production, but much more. Through them, they impact on lifeworlds, lifespaces, power relations and societal structures, and of course, on the social environment and cultural life. Information technology networking drives possibilities for changes in the fundamental categories of human life, namely, space and time.

The obsolete technologies have unfolded their sphere of activity only on the place of their existence. Computers may have the function of a chapel. The entrance to the chapel leads to the same remaining metaphysical spheres. (R.Alton-Scheidl)[5]. Therefore the development of technology is by no means limited to the information-processing professions. Furthermore it's a particular technology which has its relevance in all spheres of human activity and productivity, and therefore is changing in a remarkable way the space-time life samples. What are in particular the myths about telecommunication. A common created understanding is the saga about the home office, home work in electronic houses, etc. M. Castells[6] writes 'we are told, for example, that telecommunications allows work at home in "electronic cottages", while forms become entirely footloose in their location, freed in their operations by the flexibility of information systems and by the density and speed of the transportation network; or, that people can stay at home, and yet be both open to an entire world of images, sounds and communication flows, and potentially interactive, thus superseding the need for cities as we have known them until the coming of the information age'. Historical optimism and moralistic pessimism both convey in different tones an equally simplistic message of technological determinism, be it the liberation of the individual from the constraints of the locale, or the alienation of social life disintegrating in the anonymity of suburban sprawl. In fact, none of these prophecies stands up to the most elementary confrontation with actual observation of social trends. Telecommunications is reinforcing the commanding role of major business concentrations around the world. Salaried work at home in effect means mainly sweated labor in the garment industry while 'telecommuting' is practiced by a negligible fraction of workers e.g. in the US. Intensely urban Paris is the success story for the use of home-based telematic systems, while the American equivalent of the French Minitel completely failed to attract customers in the Los Angeles area, the ultimate suburban frontier. High-technology industries are key ingredients of new economic growth in some regions, but are unable to generate a developmental dynamic in other contexts. Societies and economies stubbornly resist being molded by the application of new technologies: In fact, they mold the technologies, selecting their patterns of diffusion, modifying their uses, orienting their functions. New information technologies do have a fundamental impact on societies, and therefore on cities and regions, but their effects vary according to their interaction with the economic, social, political, and cultural processes that shape the production and use of the new technological medium (Castells)[7].

Migration

Basically, two main streams are useful in the migration debate, the internal migration movement, (i.e., the regional flows to the accumulative centers) and the external mobilities, (i.e., from one society to another). In the classical argument, labor movements from the south went mainly to the north. The industrialized north created labor and living conditions for huge potentials of labor power, mobilizing them to the assembly lines of production. Taylorism became a key issue for the efficient way of production. Specialization and discipline were integral parts of the concept. Ford, as the successor of Taylor, stated in his classical book, that:

> We expect our men to do what they are told. The organization is so highly specialized and one part is so dependent upon another that we could not for a moment consider allowing men to have their own way (Ford)[8].

In Europe, the migration issue became a central theme, when in the eighties the idea of multi-culturalism was reinvented as a further step for the welfare state. Any discussion about migration and multi-culturalism in different European countries is to be situated in the tension field of migration, nation state and welfare state (M. Bommes and F. Olaf Radtke[9]). A further argument is that external migration can be classified into push and pull factors. In the latter the size and structure are from the indigenous labor force. Increases in full-time education postponed the entry of young people into the labor market creating a gap for unskilled migrants while indigenous workers also moved into better-paid, white-collar skilled occupations. Push factors are the unemployment, poverty and underdevelopment of labor-exporting countries, which have high rates of population increase, high unemployment and low per capita incomes. Finally, migration, does not provide economic development for poor regions which remain underdeveloped and dependent on the centers of industrial capitalism (Abercrombie et al.)[10]. Furthermore, local and external migration is strongly connected with multi-culturalism. Multi-culturalism emerged as a social phenomenon at the beginning of this century, and modern or, even better, post-modern societies have developed a political and economic structure where the immigrants and ethnic minorities play an important role. Several projects[11] in UK, Austria, Switzerland, France, Germany, Belgium and the Netherlands, give an insight on the diversity of experience in different cities and different countries.

Global Village

M. McLuhan's Global Village[12] statement, is that the periphery has a less important role to play and only the word village can be used as a metaphor in a symbolic way for the telematic connection of economic and central spaces in cities. These central spaces possess not only large enterprises with structures which can implement in a very efficient way communication technology, but at the same time, possess also a vast production capacity, which are a necessity for the use of market potentials, based on telematics. Big enterprises in industrial regions are increasing their market shares on the cost of Small and Medium Size Enterprises (SME), in the rural areas. Teleports are spectacular examples of the conversion and development of telecommunications technology. The meaning of this is of significance, due to the fact that these intelligent buildings are not an avoidable factor in modern city planning. The question of what a 'Global Village' looks like and where it lies, implies also the question: who belongs to the Global Village and who is the owner of it? Who can exploit the immense possibilities of the network? Can it be a small minority? Does this data highway posses also some other pathways, or, finally, will the existing social and regional disparities through the access to technology, be entirely frozen and will information affluence and information poverty perpetuate this condition? Of course the problem is not easy. The expansion and interweaving of markets for more services, are in favor of capital strong suppliers and demanders. The fact that the introduction of telecommunications convey in a strong way and threaten the existing local economic relations, is more than obvious. S. Sassen[13] stated in her work that:

the term Global City, which deals with the reproduction of the suburban areas where the changing changes for profit and relocation, is a day in and a day out trade. She put it in a very radical way, how the spiritual potential and production are split and with the assistance of communication technology then again unified. Equally, the existing societal structures and the solidarity within the society, are disappearing and a growing individualism is emerging, within the idea of the global city. National and global markets, as well as globally integrated operations, require central places where the work of globalization gets done. Furthermore, information industries require a vast physical infrastructure containing strategic nodes with a hyper concentration of facilities. Finally, even the most advanced information industries have a production process.

The idea of the global city is more than obvious. Some distinctions are emerging in the case of the wired city, Castels[14] writes 'the wired city is line networks, but through a variety of telecommunications technologies that lend extraordinary versatility to the system. In addition, it is a business-oriented "wired city", rather than one focused on the "electronic home"'. On the other hand, Mitchell Moss[15], has argued that 'the emerging telecommunications infrastructure is an overwhelmingly urban-based phenomenon'.

Although most discussions of new communication technologies emphasize the opportunities presented for decentralization, large cities are the hubs of the new telecommunications systems in the U.S. and are the sites for the most advanced applications of information technology. Through the revolution in communications, such global cities should be in the position to overtake and create the so-called transactional activities over national and global markets. Furthermore, they exercise power over them and they concentrate themselves exactly where the demand on such technologies emerging. The biggest part which deals with communication is still city-internal. The expansion of telecommunication and the 'Data Highway' are creating one of the main technological factors, which may change the picture and the images of the future city. As in the past, the infrastructural traffic developments have emerged new mainframes as Channels, Highways, Streets, air-traffic, etc. The main distinction now, between the above-mentioned processes is that, the new era cannot be seen and processed in an isolated focus and in particular in a certain country or city, but in contradiction to that in a worldwide global frame. Above that, of course, and in order to understand these new dimensions and their cultural reasoning, which are building a necessity and prerequisite for the development and realization of these visions.

The integration of new social environments and the lack of cultural identity are the basic obstacles for a prosperous city development.

Regional Development

One further issue is the argument that regional development is strongly connected with theories of comparative advantage and trade, cumulative causation, growth poles and centers, entrepreneurship and creative regions, stages, waves, and cycles are consistent with observed imbalances in economic development among and within countries. Seven types of interventions are commonly used by industrialized countries to affect the regional distribution of economic activity, consistent in different ways with theory: subsidizing capital and/or labor, investing in specific large-scale projects, enhancing the development of propulsive industries, improving the entrepreneurial milieu, facilitating international partnering, developing communications and/or transportation networks, and developing knowledge infrastructure.

Conclusion

In conclusion, five points build up the basic arguments:

(1)   Cities are gaining more importance, due to the decreasing significance of the national states.
(2)   The added importance to the cities is linked with their regions: Cities and the areas around them, can not be split up in a clear way, anymore.
(3)   In particular, big cities will not compete any more with their neighboring communities but only with other big cities.
(4)   The economic competitive mechanism will be extended through new samples of global co-operation and local manifesto from ideas, information etc. this interaction of forces, showed the critical factor for the future of cities.
(5)   The transition from a political to a post-political regulation form will gain more and more power. Cities and their administrations will be the centre of this development.

Notes and References

1   see also Abercrombie, N. (1994), et al., *Dictionary of Sociology*, Penguin, p.443.
2   Mitchell D.G. (1968), *A hundred years of sociology*, Duckworth, London, p.98.

3   see Abercrombie et al., p.442.

4   Rex J. and Y. Samad (1996), *Multiculturalism and Political Integration in Birmingham and Bradford*, Innovation, Vol.9, No.1,1996., p.11.

5   Alton-Scheidl R. et al., (1993), *Technologische Kultur*; Wien, p. 24.

6   Castels M, (1989), *The Informational City*, Basil Blackwell, Oxford. pp. 1-2.

7   Castels M, (1989), *The Informational City*, Basil Blackwell, Oxford. pp. 1-2.

8   Ford Henry, (1923), *My Life and Work*, London, Heinemann, p.11.

9   M.Bommes and F.Olaf Radtke, (1996), *Migration into Big Cities and Small Towns - An Uneven Process with Limited Need for Multiculturalism*, p.75.

10   Abecrombie, N. et al., (1994), *Dictionary of Sociology*, Penguin, London, pp. 156-157.

11   see for more information, *Innovation*, (1996), *The European Journal of Social Sciences*, Vol. 9, Number 1, March 1996, Multiculturalism and Political Integration in European Cities, Issue Editor, John Rex.

12   Fiedlschuster, (1991), *Die telematische Gesellschaft*, in: REGionalentwicklUNG, 2/91, p. 15.

13   S. Sassen, (1994), *Cities in a World Economy*, Pine Forge Press, 1994, p. 1.

14   Castels M, (1989), *The Infomational City*, Basil Blackwell, Oxford.

15   Mitchell Moss, (1986), Telecommunications and the Future of Cities, in *Land Development Studies*, 3 (1986), pp. 33-44, Paper delivered at the annual meeting of the International Institute of Communications, Edinburgh.

6   Technological Change and Employment in Society

Introduction

The decision to adopt more advanced technology has a significant impact on several levels within organizations. The question is whether a company when adopting new technology does so to update its production or because there is pressure to innovate. This approach by contrast, could easily lead to the idea that the market, finally could gained standardization and affect parallel reactions. The core of this chapter will throw light on the relationship between technology and inter-organizational relations with respect to innovations that bring about drastic changes in routine working procedures. While studies and surveys of the implementation of information technology are focus on either positive or negative impacts, this paper will investigate some innovative ideas and approaches offering a template which is based on group dynamics and therefore system eminent.

This chapter, however, after briefly identifying a number of different theoretical approaches to the issue, outlines a framework with in which it is possible to place empirical studies and provides a template for them. It then reports on some of the key empirical studies investigating the relation between the application of new technology and employment. It also provides some useful insights into the debate on the effects of technical change on employment. Furthermore, technology and its control will force people to rethink old models. If this premiss is true then it is socially constructed. If technology finally is socially constructed then it can be discarded. The passage from shaping to social values via standardization, form the major field of new research for social sciences.

The Technological Idea

The changing organizational structures, for instance, due to the new technological approach prove that a paradigm change of the conception of enterprises structure is underway. As in the past, ideas and approaches like the social shaping of technology, or technological determinism and finally innovations are taken into consideration. Organizational implications were always a characteristic of the adoption of technology. A certain thesis which was put forward by Gruppelaar (1988:175) stated when focussing on the technological involvement aspect, that:

> In pre-or non-modern (Western) societies technological facts have a limited context because they are separated with respect to each other and an 'un'-limited context because they are part of a non technological context. It is therefore possible that techniques that change architecture, do not change agricultural technologies are locked up in their application. These technologies are not part of the development of technology. New techniques develop architecture, not technology.

Further focussing on technological commitment, he argued that a systematic development of technology presupposes a certain autonomy of technology, a technological context. Within this context the autonomous power of technology is invented, explored and developed. Technology emancipates as 'plain pouvoir' and forces other developments. All kinds of facts appear within a technological grid. Technology itself decides what is validated as fact, it controls the pertinence and impertinence of what occurs. Technology defines the human situation and, nowadays, perhaps even the human condition. Because of this specific quality, and therefore quantity, of the influence of technology on modern society, one can say that this society has become committed to the development of technology. It is said that in modern times the pace of change accelerates, but it is the necessity of change that is decisive not the multitude of changes. The password of modern society is innovation. There is permanent pressure of innovating forces on the orthodoxy of this society (Gruppelaar 176:183). Finally, Gruppelaar, argued that this development of technology should be studied as a cultural-historical phenomenon; because of the fact that a technological development is culturally disposed, its objective power is an ethical problem. Therefore, studies on technology should concentrate on its cultural dimension. He continues by pointing out that technology is bound to cultural values and ideas, and many other ranges have been suppressed by the development of technology; the revaluation of other human possibilities and capacities will help solve problems that are now unjustly treated as technological problems. Reflecting upon such considerations, a term which deals with values and ideas is trust. Informal procedures and building groups between groups, seems that has a significant meaning in the age of the first computing restructuring. The ability to link different organizational units with decisions and persons, goes definitively back to the time of the first installation of information systems. These first structures which integrate computers in banks, for instance, created at the same time, new hierarchies within the firm and enable communication. It helped, among other things to explored trust and its role in the motives and authorities which plays in decision structuring. Recently, sociologists and economists have begun to focus more directly on the role of trust in organizations. Gambetta (1988), Granovetter (1985) and Bradach and Eccles (1989), for instance argued that trust plays an important role in facilitating economic transactions, especially in transactions embedded in social relations (Kraner/Tyler :332).

Theoretical Considerations

Technology has become proceeded to a decisional factor of societal action. Technology has created many things, which technicians disregard despite the fact that they, themselves, are the creators of such structures. However, the argument is: why can technological development be seen as an independent action and, vice versa, societal cultural action as a dependent issue, since both belong to totally different fields of societal reality?

The usual mode of scientific concepts uses theories to describe reality. Nevertheless, the quality of these theories always presents a problem. According to Willke who argued in 1989 that if these theories were measured by their usage then we would have more or less second hand theories. It is obvious that upon the conception of a theory, it is easier to measure and evaluate. Willkes (1989:10) argued that during the periods of Marxism and Neo-Marxism, the relationship between capital and work was the main problem; as far as phenomenology is concerned, it is the problem of daily life. The critical theory was here liable to the problem of emancipation, and the theory of action to the problem of social action. He thought that it was important to be aware of which theories were in vogue, and which ones had to remain in the background. Many theories which have survived the tests of time are measured by Popper's criteria, namely that theories should allow the construction, examination and falsification of hypotheses.

Glaser and Straus (1967:3) mentioned that the interrelated jobs of theory in sociology are (1) to enable prediction and explanation of behavior; (2) to be useful in theoretical advance in sociology; (3) to be usable in practical applications-prediction and explanation should be able to give the practitioner understanding and some control of situations; (4) to provide a perspective on behavior-a stance to be taken toward data; and (5) to guide and provide a style for research on particular areas of behavior.

Thus theory in sociology is a strategy for handling data in research, providing modes of conceptualization for describing and explaining. Adorno (1972:83) stated, finally that 'the application of theory remained uninfluenced by the examining practice. Furthermore theory and empiricism cannot enter the same continuum'. According to the empiricism of technological development processes, Hochgerner (1986:11) further, stated that it was usual in sociology to take technical equipment and facts into account almost exclusively as societal external factors. Few exceptions existed outside the dominant development lines of the discipline. The systematic consideration of technical aspects within social facts, the observation of technology as a societal indigenously produced element, implies a transformation of the structures and modes of operation of social relations on a long term basis. Bearing these points in mind, should the issue, the tasks, the theoretical and methodological points of sociology be extended and partly revived on this foundation? In addition, what happens with complex positions to which a medium range theory cannot offer satisfactory solutions. At this point, a mixed theory is required which can neither be postulated in a systemic theoretical form nor be gained through the state of the social theory. The general call for a system theory (Miller 1978, Bertanlanffy 1979) has become stronger than ever. D. Edge (1995:15) for instance, identified eight types of social influence on technological change: geographical, environmental and resource factors, scientific advance, pre-existing technology, market processes, industrial relations concerns, other aspects of organizational structures, state institutions and the international system of states, gender divisions and cultural factors. He also points out that this kind of distinction shows two main weaknesses: (1) Discussion of each type of influence tends to exclude the discussion of the others. More integration is needed. (2) Of equal concern, almost all existing literature lacks a satisfactory conceptualization of the precise way in which new technological knowledge emerges and evolves over time (Edge 1995:16).

Technological Changes

The advances in technology and organization recreate the manufacturing and management system and exclude the non-compatible structures by human manpower and organization. To question concepts on how organizations are functioning now, becomes not only a scholastic term for effective management but for the first time shows serious implications on the practical daily use of computer supported technology. Implementation of technology, requires more skilled workers and judgement, initiates new job qualifications which are based on technological innovations, eliminating, on the other hand, the non up-dated compatible human capital.

The rapid technological changes and their applications in many sectors of the economy, demand and create qualifications which can be extended over the life cycles of hardware and software. Furthermore, qualifications which emerge among the administration of processes are system eminent. How can new knowledge both be adopted and be trained. What is the current state of these qualifications which are shorter than the life cycles of hard and software? What can be the common point between profit oriented organizations and those which are offering social services as part of the welfare state, such as hospitals? etc. The Commission of the European Communities, focussing on this theme, published in 1992 a report on the role of technology in prolonging the independence of the elderly in the community care context (*Technology and the Elderly*, doc Eur 14419 Report 1992:114). The report has highlighted the needs of the elderly living in the community and the main characteristics of the systems of community care which have evolved in Europe to support these needs.

Definitions and Framework

The necessity of an organizational model if we follow R. Walton (1987:17) is that:

> Innovation cannot be effective unless it is guided by a vision made manifest in a model. A model is a general concept of the future organization and evolves from an understanding of the limitations of traditional organization and experimentation with alternatives. The model must be strong. That means it takes into account the fact that policies often affect other policies. An organization cannot work well unless all its policies, technical systems, and structural features are adequately aligned.

The social power of the microsystems, however, seems to be an important issue, since new organizational structures still incorporate the whole view of the market, technology and personnel, so that strategically competition can build up further. The transformation of an enterprise strategy is successful within the human component. Consequently the human function is responsible for success and establishes one of the important sources of enterprise strategy. An organization is compatible or successful, when it is not static or, if we go back to K. Türk (1989), organizations are only rudimentary on a long term stable, much more there are permanently in motion; they never reach their surface stability, neither by balance nor static, but only by movement. New organizational images produce a constantly changing social reality because their duties, due to the changing market variables, must be redefined every time. On the other hand, these processes create for the working class a determined effect. Each of the above aspects offers some insights into the relationship between technical change and employment yet none offers a complete account.

Speaking about the adoption of technology, much attention has been paid in the past to the so called effective implementation of technology (Majchrzak, 1988) or the meaning of local knowledge systems in organizations with advanced technology (Baba 1990). Further, approaches like skill based design lead to organizational effectiveness, and gain more interest about the aspect how to turn technologies into tools (Salzman 1992) or create symbolism in organizations as an emerging asset of establishing realities within the firm, based upon their own professional histories, personalities, value systems, etc., (Astley 1985). Furthermore, the aspect of the social construction of technological systems (Bijker et al., 1987) among technologists arises and many more people are now aware of the fact that culture, language and ideology are definitely terms of management. Uncertainty of technological systems, affected the same in organizational concepts, because organizational structures should become more organic and less mechanistic as Kolodny stated (1990). Flexibility, on the other hand, replaces efficiency as one of the key objectives of both the manufacturing and service sectors. Such initiatives have generated new organizational skills and methods as Computer Supported Co-operative Work. Egger and Hanappi, (1994) define CSCW as 'a group of people who has to accomplish a common task, the decision and therefore the bargaining process is limited to task-time-person assignments'. Contrary to pure process-automation where working people are treated as resources being assigned to certain tasks, CSCW pre-supposes that group members are in a position to formulate personal preferences and at least partially, can realize them. The issue is very important, especially since in order to support groups by providing them with a technical system, a basic framework has to be designed, which is a representation of the group's setting. Implementing a group decision support system therefore means building a model of the decision-making situation. Since bargaining over work distribution has a rather strong social dimension and requires the consideration of many different planning parameters its support by technical features seems to be justified.

Since different innovative products have found their way to the market, large enterprises react with the same technical and strategic policy and are biased on rudimentary organization models. The uncertainty of market it maintains itself and a non static qualification wave from the large enterprises moves down to the small and medium size enterprises.

Other and forthcoming models on an information system structured economy argue that the buyer is also buying a piece of culture in his enterprise. From that point of view, small firms are more dependent on highly paid manpower with advanced qualifications-who may have gained these advances and further knowledge in large enterprises. A new realization has dawned with respect to the transferable learning processes within the firm and within the international transfer of technology. On the other hand Taylorism as the common classical strategy by the implementation of work in the industry, can not be justified (Wobbe 1986).

Reflecting upon such considerations further other ideas, which are in the debate now gain efficiency and define consequences in working structures. Empowerment has become, for instance, a popular notion in leadership theorizing. It is based on the idea that, given the freedom, scope and resources to achieve organizational goals, people will, in effect lead themselves - if it is in their interests to do so. Leaders, therefore, do not tell others what to do, or attempt to sell their ideas to them. Rather, the leader's role is to help others achieve their own ends creatively by helping them to discover their own potential, and clearing a pathway for them. The leader, in this way, gives power, to his or her followers. The leader is a facilitator of other peoples action. Empowerment is an extension of democratization in management and the fading of the authoritarian leader (Srivastra 1986).

Technology and Employment

As M. Campbell, (1994) pointed out, from a case study in the UK:

> In most cases the focus is on capitalistic or, at least, non-democratic forms of organization where the deployment and exploitation of the technology is purposive and has the intent of achieving organizational goals either in direct form of the production and distribution of goods and services for markets or in an indirect form of facilitating and supporting the internal administrative functions of the organization. Rarely is consideration given to an examination of the purposes, process and problems involved in introducing new technologies into non-capitalist, voluntary organizations where the goals of the organization are, in theory, determined by the members of the organizations and where the leaders of the organization are, again in theory, subject to the democratic control of the members. Yet, the capacity for the new technologies to alter fundamentally both the boundaries between the technical and social relations within such organizations is surely considerable.

Campbell further stated that according to the effects on employment, the most extensive, comprehensive and recent study is reported in Christie et al., (1990) 1200 manufacturing establishments were surveyed in 1981, 1984 and 1987 with a view to establishing, inter alia, the employment effects of the introduction of microelectronics in both production processes and products.

By 1987, 75 percent of factories were using microelectronics (double the figure from 1981). We focus here on the direct effects of process innovation. Overall the ratio of job losses to gains had increased over the period from 28,000 in 1981-3 to 82,000 in 1985-87. However to put this in context, it is equivalent to just over 2 jobs per factory per year and 52 percent of factories reported no change at all in numbers employed due to microelectronic applications. Nonetheless multi-variate analysis shows that the greater the use of microelectronics the greater the job loss especially in food, drink and tobacco, chemicals, vehicles, printing and mechanical engineering sectors as well as in large firms overall. In small firms (less than 20 employees) the effect on employment was positive! Overall, on the number of job losses due to the direct effects of process innovative was very small (gains and losses) compared to the overall job losses in the factories through the 1980s. They conclude that while new technology is not the 'job killer' that some suggest, it does allow large productivity and output gains to be achieved without significant employment growth, a finding from micro analysis that parallels the aggregate study of Freeman and Soete (1987) and implies that there will be little if any growth in the level of employment in manufacturing in the future despite increases in output.

Basic aspects of the organizational and production concepts in European countries seems to differentiate from flexible specialization (FS) to lean production (LP) and human centered systems (HCS). While the HCS tend to incorporate mechanical construction, the FS is more dedicated to textile industry while LP has a validity in the car industry. European industrial policy was concentrated in the early 1990s on several main topics, which should have improved and maintained its survival in the world market. Such policies were: (1) technological modernization, including R&D, renewing and development of infrastructure in transport, energy and telecommunication, (2) regional modernization, which includes the structural, social and agricultural fonds, (3) ecological modernization which involves the Delors-II-packet, Research standards, (4) integration of markets (common market), Mediterranean countries, East Europe, and policies in competition, order and trade issues, (5) social and economic modernization which includes professional education, social dialogue, job security and a system of health protection.

Interests and Participation

Given these explanations, there is a need to confront the above issues in such a way as to know how these aspects could change. Changes in the economy cause new conditions for the inner life of organizations which includes trade unions and other relevant associations. Windolf (1989:367) pointed out that:

a former homogenous working class is disintegrated into multitude professional groups, which exist far away from each other according to their professional, economic and political interests; so that a huge bureaucratic organization can not represent them effectively anymore.

The global interests of the firm target lies on a concept, how employment operates, that means when workers start to work, finish work, contents of the transformed working actions etc. Galbraith (1972:225) argued that 'these (aspects) are not anymore to be found under the innovation and modernization effect of the free market economy' and by that Windolf emphasizes that 'trade unions belong to a specific development level of the industrial system. When this level has been superseded, trade unions lose their original powerful position'. On the other hand, participation might be taken to a new course, where it can act against the deregulation process, which was described above. Concrete steps could be the right to be informed as employee by the enterprise and by that allowing the unions on a local basis to act effectively. In general an extension of participation of working organizations, creation of work and technology and protection against unlawful dismissal could be the first steps.

Conclusion

It is obvious that the creation, invention and implementation of technology, sets a conditional action which in some cases is characterized by determined decisions. Effects on other groups and organizational work lead to the necessity of asking for parameters which are based on accurate information. Hanappi and Egger (1996:454) stated that modeling bargaining situations leads to the necessity of considering existing decision theories and of assessing them in respect of their applicability to working groups. Further, it seems that the technical support of group decision making-such as bargaining over work distribution -seems to become a more and more important question. Telework, virtual organization and borderless offices are only some examples of future visions which will require adequate models of planning procedures of working groups. From a system designer's point of view bargaining is a multilayered social action.

The effects of new technology, its reproduction and its adaptation in the production process affect various sectors of society. One main effect is the so called societal deregulation of labor relations. At this point various interests act on the sphere of the employer's activities. On the other hand the permanent reorganization of labor sets as its real goal the circulation and accumulation of capital, and the perpetuation of the existing societal structures. The technical systemic structure of work was not planned at the beginning of the technical intention nor in the previous forms of work organization.

Wobbe (1986:37) stated that:

> The adaptation of new technology was mainly affected in increasing productivity, cost saving and work saving. This is finally the core of the capitalistic-enterprise rationalization. If the production level remains the same in the enterprises then this will lead to the reduction of labor posts.

Out of that two possible ways can be seen of management concepts due to using power of labor. The first *wave* is totally technologically oriented and the management invests only in this direction, using labor power to assist the machines. In addition they demand strict control and centralization of decision making.

Using the term revolution J. R. Beniger in his book *The Control Revolution*, stated the following regarding control and revolution:

> Revolution, a term borrowed from astronomy, first appeared in political discourse in seventeenth-century England, where it described the restoration of a previous form of government. Not until the French Revolution did the word acquire its currently popular and opposite meaning, that of abrupt and often violent change. As used here in Control Revolution, the term is intended to have both of these opposite connotations. Beginning most noticeably in the United States in the late nineteenth century, the Control Revolution was certainly a dramatic if not abrupt discontinuity in technological advance. Indeed, even the word revolution seems barely adequate to describe the development, within the span of a single lifetime, of virtually all of the basic communication technologies still in use a century later: photography and telegraphy (1830s), rotary power printing (1840s) the typewriter (1860s), transatlantic cable (1866), telephone (1876), motion pictures (1894), wireless telegraphy (1895), magnetic tape recording (1899), radio (1906), and television (1923) (1986:70).

The second wave is based on an empirical unideological concept. Here the management discerns the weak structures in technology and invests in human competency in connection with technology and accepts local decentralized decisions. Both systems are based of course on criteria of economic efficiency, thus the formation of the established industrial society delivers the framework for such decisions. Interests which reflect these concepts could be realized in the environment of the firm. On the other hand the technical systemizing of work in enterprises is to be seen as a result of these preconditions on the level of macro structures. In a model where the perspective of a logical construction of enterprises or organizations will conclude with their extrovertial and introvertial perspectives could in this sense analyze the more and less successful structures. Such structures show not only various dynamic aspects but they find themselves subject to a continuing change.

A further step in the debate about technology, has been taken, and new approaches are emerging considering the role of design in the shaping of technology. Two workshops within the action A3, (Management and New Technology) and A4, (Social Shaping of Technology), have taken place and the target was to identify reasons for the technical and social dimensions of artefacts which are built at the same time.

Appendix

Campbell M. and Baldwin S. (1992) in their work *Recruitment Difficulties and Skill Shortages*. A Report for the Employment Department, Policy Research Unit, stated that the debate about technical change in relation to new or information technology began in the late 1970s. Whilst it is not possible to review all these studies here it is important to draw attention to the range of studies to see how they fit into our framework outlined above. Jerkins and Sherman (1979) estimated that the introduction of IT could reduce employment by 5 million by 2003: equivalent to a 23 percent reduction in the employed labor force. They also predicted unemployment rising to 15 percent by 1985. These estimates focus entirely on the widespread and rapid introduction of IT in the process of production in manufacturing and services and wholly on its direct effects. Batstone and Gourlay (1986) in another survey found that half of the respondents reported a 'negligible' effect of IT on employment, whilst 1 in 4 reported that it had been associated with a loss of jobs of 10 percent or more. Wobbe (1989), in a much more specific and detailed study of the effects of robotics on employment, estimated the direct effects at a loss of 120,000 between 1980-95, but total compensation effects of all the other effects at 90,000, giving a net job loss of 30,000 or 0.8 of a person for each robot introduced.

In a study very much in the structuralist mold, Freeman and Soete (1987) developed a sophisticated analysis based on their TEMPO research programme. Suffice it to say that they concluded that 'new technologies could play a very important part in generating new employment opportunities' and that the combined effects of the compensation mechanisms (productivity, prices, output, income, new products and so on) swamp the direct effects of labor saving process innovations. Furthermore they showed that the employment problem arises not from too rapid a growth of productivity, occasioned by the application of new technology to process innovations, but from too slow a growth of productivity occasioned by a lack of process innovation. Nevertheless this does not imply that they adopt a free market approach, for the latter has occurred precisely because of market failure, the uneven diffusion of innovations and limited product innovation, low rates of growth of aggregate demand, inappropriate government policies and a tendency to undertake process innovations for rationalizing (restructuring to reduce unit labor costs and capital deepening) rather than expansionary (capital/widening and increases in output) purposes. They also found that the only manufacturing industries to increase employment were precisely the microelectronics based sectors. However they did find a significant 'break' in the relation between output and productivity growth (the so called Verdoorn relation) in the mid 1960s; so that whereas previously a 2 percent rate of growth in output was required to maintain employment, in the late 1970s and early 1980s this had grown to 3.5 percent, thus implying a shift in the balance of the effects of process innovation from output to direct effects.

References

Adorno, Theodor. (1972), *Soziologie und empirische Forschung,* in: Adorno et al., *Der Positivismusstreit in der deutschen Soziologie*, Darmstadt und Neuwied, pp. 81-101.

Astley, W.C. (1985), *Administrative Science as socially constructed truth*, Administrative Science Quarterly, 30, 497-513.

Baba, L. M., (1990), *Local Knowledge Systems in Advanced Technology Organizations*, in: L.R. Comez-Mejia, M. W. Lawless, Organizational Issues, JAI Press, Connecticut.

Batstone E. and Gourley S. (1986), *Unions, Unemployment and Innovation.*, Blackwell,

Beninger, J. R. (1986), *The Control Revolution*, Harvard University Press.

Bertalanffy, Ludwig. (1978), *General Systems Theory*, 6th revised edition, New York.

Bradach, J.L. and Eccles, R.G.(1989), *Price, authority, and trust: From ideal typs to plural forms.* Annual Review of Sociology, 15, pp.97-118.

Curtis, B., Herb Krasner, Neil Iscoe.: (1988), *A field study of the software design process for large systems,* Communications of the ACM, Vo. 31, Nr. 11. pp 1268-1287.

Campbell M. and Baldwin S. (1992), *Recruitment Difficulties and Skill Shortages,* a report for the Employment Department, Policy Research

Unit.

Christie, I. et al., (1990), *Employment Effects of New Technology in Manufacturing,* Policy Studies Institute.

Downey, H.K. and Arthur P. Brief.: (1986), *How Cognitive Structures Affect Organizational Design:* Implicit Theories of Organizing. in: Henry P. Sims, Jr., Dennis A. Gioia and Associates: The Thinking Organization: Dynamics of Organizational Cognition. pp. 165-190. Jossey-Bass Publisher.

Edge, D, (1995), *The Social Shaping of Technology,* in: Information Technology and Society, A Reader, Edited by N.Heap et al. Open University, Sage.

Egger, E., Hanappi, H. (1994), *Multi-criteria decision making in groups: A game theoretic model.* Paper presented at the international symposium on Human Interaction with Complex Systems Greensboro.

Freeman, C. and Soete, L. (1987), *Technical Change and Full Employment,* Blackwell., UK.

Gambetta, D. (1988), *Can we trust trust?* In D. Gambetta (ed.), Trust: Making and breaking cooperative relations UK:Basil Blackwell, pp. 213-238.

Glaser, B./Straus, A. (1967), *The Discovery of Grounded Theory,* Aldine.

Granovetter, M.(1985), Economic action and social structure:The problem of embeddedness. American Journal of Sociology, 91,pp. 481-510.

Gruppelaar, J. and Muskens, G. (1988), *Global Telecomunication Networks: Strategic Considerations,* Kluwer, p.175.

Hartmann, Michael. (1986), *Strategien und Resultate der Verwaltungs-rationaliserung,* in: Journal für Sozialforschung, Heft 2, p.180.

Hochgerner, Josef. (1986), *Arbeit und Technik. Einführung in die Techniksoziologie,* Kohlhammer Verlag. Stuttgart.

Hörning, Karl. (1989), *Vom Umgang mit den Dingen,* in: Weingart, Peter. (ed.) Technik als sozialer Prozeß, Suhrkamp. Frankfurt am Mai, p. 90

Gioia, A. D. (1986), *Symbols, Scripts, and Sensemaking: Greating Meaning in the Organizational Experience.* in: Henry P. Sims, Jr., Dennis A. Gioia and Associates: The Thinking Organization: Dynamics of Organizational Cognition. pp. 49-74. Jossey-Bass Publishers.

Keen, P.G. (1990), *Telecommunications and Organisational Choise,* in: Fulk, J. Steinfield, Ch. *Organisations and Communication Technology,* Sage Publications, London, pp. 298-301.

Kolotny, H. F., (1990), *Some Characteristics of Organisational Designs in New/High Technology Firms,* in: L.R. Comez-Mejia, M. W. Lawless, Organizational Issues, JAI Press, Connecticut.

Kraner, R. Tyler, T (1995),*Trust in Organizations.* Sage, p.332.

Marcuse, Herbert. (1941), *Some Social Implications of Modern Technology,* in:

Zeitschrift fuer Sozialforschung, Jg. 9/1941, Nr. 3/41, pp. 414-439.

Miller, James. (1978), *Living Systems,* New York.

Nakos, Pythagoras. (1990), *Informatik und Human Resources Management,* in: Informatik in der Verantwortung der Unternehmensführung, Institut für Informatik der Universität Zürich, Nr. 990.02. p.135.

Pouvourville G. (1985), *Hospital System Management in France and Canada:* National Pluralism and Provincial Centralism, in: Sc.Sci. Med. Vol. 20, No. 2. pp. 153-166, Pergamon Press.

Srivastra, S.(1986), *Executive power.* San Francisco, Jossey-Bass.

Star, Leigh, S.(1989), *Layered Space, formal representations and long-distance Control: the politics of information.* Fundamenta Scientiae, Vol. 10, no. 2, pp. 125-155, Brazil.

Trevino, L. K. Robert H. Lengel, Richard L. Daft.: (1987), *Media Symbolism, Media Richness, and Media Choice in Organizations,* A Symbolic Interactionist Perspective. Communication Research, Vol.114 No. 5, October 1987, pp. 553-574, Sage Publications, Inc.

Türk, Klaus. (1989), *Entwicklungen in der Organisationsforschung,* Ferdinand Enke Verlag. Stuttgart.

Walton, Richard, (1987), *Innovating to Compete,* Jossey-Bass, CA.

Willke, Helmut, (1989), *Systemtheorie entwickelter Gesellschaften,* Juventa, München.

Windolf, Paul. (1989), *Vom Korporatismus zur Deregulierung,* Journal für Sozialforschung, Heft 4. p. 367.

Wobbe, Werner. (1977), *Menschen und Chips,* Sovec Verlag, Göttingen. p.37.

Weick E. K. (1990), *Technology as Equivoque: Sensemaking in New Technologies.* in: Goodman/L.S. Sproull and Ass: Technology and Organizations, Jossey-Bass, San Francisco, Oxford pp. 1-44.

7 Structuring Technology and Working Environments

Social Processes and Engineering

As has been pointed out in previous works (Katsikides 1997), an effort has to be made to understand societal transitions and methodological means, or, as Talcott Parsons advocated, using sociology to study the relationship of an individual's experience to society and history, the starting point for the sociology of technology must be through science. A key point, which was made by Hochgerner et al., (1994), refers to the 'formative principles' which will create both a source and a framework. Formative principles as theoretical concept cannot offer a normative, fixed and true picture of societal developments, as this concept was developed and is strongest in industrial countries where the principles of hierarchies, objectivism, and growth are at the heart of the theoretical frameworks (and subsequent empirical analysis). However these signs do not pretend to show the direction that development will take, nor do they describe normative expectations in reference to the historic dynamics of civilisation. Rather, they represent a practical implementation and, in large social systems, a specifically related further development of the micro-sociological term figuration. Included in this term are transformation models in which individual's act according to the situation, 'not only using their intellect, but also with their complete body and their every action in their relationship to each other (Elias 1978:42). Furthermore, Hochgerner et al., have explained why hierarchies, objectivism and growth are termed as formative. Claiming that they are intrinsically valid because they each comprise both material and idealistic elements and organizing social facts (laws, customs and traditions, roles, expectations, etc.) such that the societal development they support is in fact secured by their existence. The security of this continued existence allows for remodeling and changes (throughout the society and even concerning its formative principles). For industrial societies which recognize 'growth' as a constitutive necessity, constant change could even guarantee preservation. Such a societal development can only be maintained continuously if it is able to remodel itself by adopting to constant change in a controlled way. This regulating mechanism which controls human behavior according to the specific needs of a certain given societal development, is termed a formative principle. It organizes existence and change in social behavior over several historical eras without itself being restricted to the respective form of that time. On the other hand, objectivism deals with human behavior when this is standardized, or 'functioning'. Finally, the study of what has come to be termed as the sociology of technology does in fact incorporate elements of sociological methods, since they can illustrate social behavior in a regulated societal system where technology and formal foundations create the context and the perspective. In the above mentioned work (Katsikides 1997) the aim was to show that there is a variety of theoretical issues which can be directed to mainstream sociology of technology. One common understanding which derives from the research in the field reveals that most sociological studies on technology use the comparative method, and the remaining apply to the field of technology assessment. We have argued that technology reflects the synergy of power and societal processes, and these must be analyzed under the foci of sociology of science or even of the emerging sociology of information.   While sociology of information should address a variety of theoretical perspectives that can be directed towards the social phenomenon of information, they alone do not give sufficient insight into the nature of information either as an object of disciplinary discourse or as an object of nature (Balnaves 1993:108). The emerging approach is that an entirely new concept is required and that there is a vital need for improved analysis with respect to the assessment of technological issues. It can be argued that theoretical considerations have to be linked with practical methodology in order to evaluate technological and societal approaches, because different sets of complexities exist between the cultural and the operational aspects of the functional role of technology. However the issue here is more complex, and the argument can be summarized as follows. The first problem relates to methodology, where it is clear that a global approach, whether theoretical or empirical, reaches its limits very quickly. The second problem is a more general issue that refers to all the social sciences: a common direction to resolve common social phenomena is lacking. Thirdly, it can be argued that a new approach is needed, which would focus on a detailed evaluation and provide a synthesis of all the intervening variables involved in the technological discussion. One example of such an approach is the ARS model (Katsikides 1994). Finally, technological developments, like other social, economic, and technical approaches, are not socially neutral, and in the end they deal with different traditions (European, US, Scandinavian, Japanese, etc.). As such they accumulate social processes and reflect them, or, as Thomas Kuhn (1970) put it 'a failure to assimilate fully new conditions and technology will strain the existing structures' of society'.

Moving now forward in order to analyze the position and values of technical workers, several ideas and notions could be made. S. Crawford, for example, in his work 'Technical Workers in an Advanced Society', which has been published by the Cambridge University Press in 1988, has pointed out that only two empirical studies compare the position and values of technical workers in old and new industry. Zussman's (1985) book on American engineers and Whalley's (1986) volume on British engineers. Crawford introduces in his above mentioned book the study of French technical workers both complements and builds on their important work. He positioned that the findings of an investigation into the work, careers, and ideologies of French engineers and managers employed in two industrial settings, a traditional metal-working firma and an advanced telecommunications firm. What makes actually this notion so interesting is the passage from the theoretical considerations over to the real life on technological systems. The old question, who drives whom either technology or society, and vice versa, seems that it has reached a dead end. In this chapter, engineers are the object of investigation. Several empirical studies as Crawford (1988) stated, have tried to describe the largely unsuccessful struggles of American mechanical, electrical and other engineers to develop codes of ethics, unify, and form an association powerful enough to effectively represent their interests of employers and the state. It must be taken into consideration that other aspects of the advanced society and its industrial system are construct the main protagonists in this issue. Trade Unions, for instance, education and training, collective agreements, and market performance, play a vital role on technological decisions. Contemporary status and identity of engineers, was the aim of the work of Durand (1972). Some of these works were, Calhoun (1960) in his book *The American Civil Engineer* or Layton (1971) in his work *The revolt of Engineers*, Noble (1977) in his pioneering book *America by Design* or Meiksings (1982) with his work *Autonomy and the Engineer*. Further studies in France as Grelon (1986) *La Modèle de l'école d'ingénieurs comme formation à la technologie et comme insertion dans la société* on the history of French engineers, analyses the historical structures. The outcomes of all these works were characteristically of their epoch, the idea, which was highlighted was if we can't change the industry or the capital behind it then the engineers. This conception, however, finally failed. New industries, are new technology oriented structures, our aim is to look at several industrial enterprises, where information systems, electronics and telecommunications employ 'technicians', engineers and technical workers.

Salzman et al., (1994) positioned that the main view of engineering is that it is 'applied science', that is, it is the application of scientifically and objectively determined principles. The 'scientific view' of technology is that advances in knowledge are largely independent of subjective influences. Thus, technology reflects engineers' calculations of the most economic and efficient designs to utilize that knowledge. This is the dominant view of engineering as expressed, for example, in the U.S. by the accreditation Board for Engineering and Technology definition of engineering as 'the profession in which a knowledge of the mathematical and natural sciences gained by study, experience, and practice is applied with judgment to develop ways to utilize, economically, the materials and forces of nature for the benefit of mankind'. Insofar as social choices or values are considered, they are regarded as important for decisions about the use or development of a technology but not as an integral part of the design process. Social scientists and historians have examined the 'social' character of technology. David Noble (1977), who contributed some of the first analyses of the social forces shaping engineering designs, challenges the idea of an inner logic driving technology development and leading to specific designs. Tracing the rise of engineering as a profession within the confines of industry (in contrast to other professions which developed as independent practices), he finds that, as a result, the values of engineers reflect those of their employers, only marginally distinct if at all, from managerial objectives. Engineering work, Noble concludes, is oriented toward developing technology that reinforces the existing political and social order.

Case Study in Austria

Little work has been done, which explores the ecological and sociological attitudes of engineers and technicians (of both sexes) in large enterprises in Austria. This research note reports on the results of a detailed, quantitative and qualitative study. The study which was conducted between 1991-1992 by a project group under J. Hochgerner at the Chamber of Labor in Vienna, Dept. of Social Sciences, and is entitled *Technicians within the technical change* (TTW-Techniker im technischen Wandel) throws some light upon the inter-organizational processes within the engineering profession. It lays out a theoretical investigation of the main lines of development of technological changes as well as the results of extensive empirical research. Based on 1042 standardized interviews with technicians from all over Austria, (65 expert interviews), 12 case studies in technically intensive industries, various analyses were made. In addition another study which has been conducted between 1994 and 1995 under J. Hochgerner and his research team at the Chamber of Labor and Vienna, Dept. of Social Sciences, which has focussed more on the internal processes and structures of the trade unions, has underlined the basic arguments of this research note. The survey concludes by considering the implications of these changes, mentioned in the study, that various internal processes, innovations and crises in enterprises have to cope with technological and political decision making for trade unions and their future in Austria.

The Context

Rethinking the way organizations exist and the advent of the business re-engineering approach are creating new research areas and questioning the traditional thinking in applied social sciences. The social impact of technology has been widely studied and a lot of attention has been paid in recent years to many concepts and various ideas and notions in this field. Effective implementation of technology, for instance, as stated by Adler and Helleloid, (1986), or the meaning of local knowledge systems in advanced technology organizations (Baba, 1990). Approaches like skill based design, which leads to organizational effectiveness are becoming more and more relevant when considering how to turn technologies into tools (Salzman, 1989). Furthermore the issues of social construction to technological systems was a breakthrough in the area by the works of Bijker et al., (1987) and have led to the understanding that culture, language and ideology are part of management and business strategies. Likewise the uncertainty of technological systems which led to the realization that organizational structures ought to become more organic and less functional. The key point, between the social impact of technology and uncertainty itself within technological systems is the engineering profession. Of course the connection between technology and society is a subject of long standing debate i.e., Ellul (1964), or Bittner (1993), or Berka/Hochgerner (1994). Salzman and Rosenthal, (1994), have identified two basic approaches: one approach is that social scientists should bring to builders of technology the body of knowledge about the social implications of their work. In this way these concerns can be integrated into the development environment that is otherwise highly focussed on goals of technological development. In this line of argument, engineers are prevailed upon to take more social responsibility for their work. Alternatively, it is argued that 'effective design' requires engineers to be divorced from such concerns. Many would argue that, provided a task that is well defined and a clear mission to b accomplished is evident, the engineer can process to create the optimal technology (see Salzman et al., 1994). The hitherto image of the specialized technician embodied the idea of securely being able to cope with possibly negative effects on the 'circumstantial' masters of technology. On the other hand, the TTW study clearly shows that the differences which are so often mentioned can not - or in any case can no longer be determined purely by professional qualifications. No essential differences exist amongst central ecological and socio-political opinions and technological constitutions which can be characteristically explained directly by the 'technological profession'. A process has existed since the mid-eighties which solves the particularities of technicians' political persuasions.

The Study

Considering the soundly safeguarded empirical results, the opinions of technicians which have never before been proven in this way concerning environmental issues are quite astounding. There upon it should be expected that firstly the image of the technician, secondly technicians' self-confidence and thirdly their participation in the decision-making processes in technology and environmental politics be essentially changed. A further question is in fact a two tier one: is the path being cleared in ecological awareness by members of the technological professions to initiate a 'turning point'- basically pushing the heavyweights of the original orientations out of the way? Secondly, could the key role played by technicians (a gatekeeper function), equally lead to a change in the path of ecological and social sense, perhaps even to 'another' technology? The answer to the first of these questions is, in the light of the following results, yes. These so called critical attitudes towards technology are nowadays just about as widely spread across the profession as across the entire population. Three quarters of the 1042 interviewed technicians believe in the 'heretical' opinion that 'technology itself is a danger, if people do not learn how to cope with it'. Seventy seven percent responded positively to the demand that wasteful packaging should be banned for all goods. Sixty nine percent wanted the collective idea to be taken more seriously into duty. 'Technicians themselves should make more of a commitment to control the consequences and the human form which technology is taking'. The comparison of the key tests suggests that in the last five years in particular a change has taken place, 8 percent of respondents suggested that they were explicitly in favor of 'rigorously saving energy' and 'dramatically reducing traffic'. Three quarters of those professional technicians themselves concerned with such measures support the causing principle, and generally support further strategies as well for the improvement of controls and themselves. These statistics and other similar ones reveal a potential for technology and also one for environmental political discussion, the value of which was hitherto underestimated and, just as importantly, was not taken seriously by technicians and engineers amongst themselves. Whilst about two thirds of the sample openly admitted to being environmentally conscious and to having accepted a critical approach with respect to technological development, most of the technicians believe their professional colleagues to be without doubt traditional thinking, optimists. Yet from those who are professionally active technicians his amounts to barely 30 percent. Although this figure lies above the average based on the findings of other corresponding representative interviewing, it points towards the fact that this profession simply has a clear minority position for the unrestricted concept of progress. Whether the second part of the question which was posed earlier can be answered equally as positively is less certain, for every honestly determined practical implementation of opinion into real behavior and the realization of alternative technological and political illusions will be opposed by numerous structural and organizational restrictions. Thus it can be said that amongst 'technicians' just like in any other profession there are those who simply 'move with the flow'. Those who tend to be inclined to react to the current trends corresponding to that specific time, and to maintain more critical positions than they can themselves actively achieve. However, a growing potential could be established which would appear to be against these obligations and more flexible towards the accompanying and corresponding comforts. Technicians, particularly the better qualified and those who will become more relevant engineers in the future increasingly with more indifference the products and consequences of their work, the traditional way of thinking and also the essential facets of their public image as well as the way in which they see themselves.

Participation

In addition to the environmental issues, a second essential outcome concerns working organizations and one of the central overriding themes of working relations, is the issue of participation. The research indicates that representatives of the technological profession are in no way unpolitically minded, and that trade unions being merely automatic uninterested problem solvers for the workers. Not only do more than half (54 percent) of them support the demand that the right to participate should be extended to the employees, a convincing majority of about two third of them proposed an essential qualitative extension in the character of industrial participation. They simply do not want merely to determine themselves concerning their undisputable individual rights and working conditions, but also about professional goals and products. They wish to be able to make essential decisions directly concerning industrial and technological politics, but also indirectly concerning environmental politics as well.

In all these areas, being an essential concluding consequence of the TTW study, it should be seen that representatives of the technologically qualified profession are apathetic towards the collective work-based responses but instead pursue individualistic extreme political ends. Such aims ensue from the context of working organizations within industry, particularly with reference to industrial and industry wide further training. Thus a revealed requirement of this specifically qualified profession exists even stronger than before, which has been produced and further developed as a result of their own ability, is able to use ecologically sound controls. The higher the formal professional qualification is, the more seriously the deficit of further training will be accepted, in which it concerns to only a minimal extent deficits in the unquestionable technological specialized qualification. The increasing demands of the economic, social and legal qualifications are being greatly outweighed and which, during the course of the first professional training, shows obvious shortcomings.

Technological Advice and Areas of Study

It may be easier to understand how the technicians profession could proceed to a mere qualitative content, it should be possess:

(1)   extended possibilities for clear expression and further development of their views of society and of environmental politics,
(2)   opportunities within industry for the perception of extended participation rights, as well as
(3)   improved entrance requirements and industrial and industry wide forms of further training.

This can be determined by means of:

(1)   the allotment of time during working hours within the industry for discussion and deliberation about work, professional targets and products and product evaluation as well as the ecological consequences of the developed products and finally about their achievements.
(2)   the proposing of new specific and cooperative forms of further training, in the form of so called 'circular studying': this proposes to be suitable instrument essentially to fulfill the appealing demands of training (specialized training embedded in primarily necessary social and economic qualifications). It has been 'imported' from Sweden and is the first pilot attempt, together with the various teaching materials for the topic 'work and technology'.
(3)   the establishing of a network of external advisers for working groups of technicians (study groups for engineers and for certain specialized

groups etc.) within the industry. As part of the current European transformation situation, a completely new research and advisory council for trade unions, industry as a whole and its workforce.

Interested representatives of the employees, just as employers, nationalized industries and lastly professional representatives of the engineers find themselves in this situation partly facing a completely new challenge. There is an good chance that technicians (engineers, specialized workers and academics) are ready in an extremely dedicated way to take part. They themselves are in the best position to be able to organize and offer technological advice. Associated institutions are being already established, to offer this likelihood (e.g. through cooperative further training based upon this model of circular studying) so that these motivations and abilities can be activated. These further implementations could in fact bring about a step in the right direction for ecological change and socially better suited technological development. Salzman, H. (1994:282) in his contribution *The Social Context of Software Design*, in: S. Katsikides, (1994, *Informatics, Organization and Society*, Oldenbourg Verlag), stated that the work in the general area of social construction of technology has been an important counterbalance to the long tradition of studies that abstract technology from its social context of creation and use. This approach is important for focussing attention on how technologies are constructed and used through interpretation by users. That is, how the 'meanings' of technology are constructed by users rather than inherent to a technology. These approaches provide important analyses of the shaping of technology within the user space. Some argue that technology is just one type of node in actor networks and, in this network, technology holds no special status or attributes via-a-vis other parts of the network. That is, the people/technology distinctions are not particularly salient. Some of the writing in the extreme social constructivist school seemingly argues that there are no salient properties of technology because all depends on social context and interpretation (e.g., Kling, 1984). Certainly the variability and inconsistency of perceptions about technology and its impacts suggest that technology is not absolutely determinant in its impact. However, at times it does seem that the background 'fuzziness' and contingency of propositions about technology and its impact become the dominant foreground for the social constructivist analyses. Instead, we argue that interpretations of technology by individuals are bounded and structured in some systematic ways that allow for analysis at the supra-individual level. In a review of German perspectives on social shaping of technology, Rammert (1992:83) writes:

The works based on the social constructivist approach have had a mixed reception among German researchers of technology. On the one hand, they have been faulted for what is perceived to be their exaggerated theoretical pretensions. The interactional level is said to be treated as absolute at the cost of the socio-structural level.on the other hand, the empirical studies and the related procedures of the social constructivist approach have encouraged proponents to expand the field of inquiry and increase their consideration of socio-cultural influences.

References

Abercrombie N, et al.(1994), *Dictionary of Sociology*, Penguin, WHERE?

Adler, P. and Helleloid, D, (1987), *Effective Implementation of Integrated CAD/CAM: A model.* In: IEEE Trans. on Engineering Management, Vol.EM-34, No. 2.

Adorno, Theodor. (1972) *Soziologie und empirische Forschung*, in: Adorno et al., der Positivismusstreit in der deutschen Soziologie, Darmstadt und Neuwied.

Baba, L. M. (1990), *Local Knowledge Systems in Advanced Technology Organizations,* in Gomez-Mejia, L.R. and Lawless, M.W. (eds) High Technology Management, Connecticut, Jai Press Inc. pp. 57-75.

Balnaves M. (1993), *The Sociology of Information*, ANZJS Vol. 29, No.1, March.

Bammé A. et al., (1983), *Maschinen-Menschen, Mensch-Maschinen, Grundrisse einer sozialen Beziehung,* Reinbek.

Basalla G, (1988), *The Evolution of Technology*, Cambridge University Press.

Bijker, W, et al., (1987), *The Social Construction of Technological Systems*, Cambridge/Mass, London.

Bittner, E. (1983), *Technique and Conduct of Life,* Social Problems, 30, pp. 249-261.

Calhoun D, (1960), *The American Civil Engineer,* MA: Harvard University Press, Cambridge.

Crawford, S. (1988), *Technical Workers in an Advanced Society*, Cambridge University Press, p. 1.

Durand M. (1972), *Professionalisation et allégeance chez les cadres et les techniciens,* Sociologie du Travail, 14.

Elias, Norbert, (1978), *Was ist Soziologie?*, München, p.42.

Ellul Jacques, 1964. *The Technological Society*,Vintage Books, New York.

Elster Jon, (1983), *Explaining Technical Change*, Cambridge University Press, Cambridge.

Gomez-Mejia, L.R. et.al., (eds), *New/High Technology Firms*, Jai Press Inc.,

Connecticut, pp. 166.

Gehlen, A, (1986), *Die Seele im technischen Zeitalter*. Sozialpsychologische Probleme in der industriellen Gesellschaft, Reinbek.

Grelon A, (1986), *Les ingénieurs de la crise*. Edition de l'école des hautes études en sciences sociales, Paris.

Habermas, J, *Technik und Wissenschaft als Ideologie*, Frankfurt am Main, Linde, 1972, 1982.

Hochgerner, Josef, (1986), *Arbeit und Technik. Einführung in die Technksoziologie*, Kohlhammer Verlag, Stuttgart.

Hochgerner, J. (1993), *Techniker im technischen Wandel*, AK-Wien.

Hochgerner, J, Berka, G, (1994), *Social Environment of Technical Progress*, in Katsikides et al., *Patterns of Social and Technological Change in Europe*, Avebury, Aldershot.

Huelsmann Heinz, 1985. *Die technologische Formation,* Berlin.

Katsikides, S. (1997), *Sociology and the Functions of Technological Automony*, in: Innovation, The European Journal of Social Sciences, Carfax. Vol.10,11, pp. 195-201.

Katsikides, S, et al. (1994) *Patterns of Social and Technological Change in Europe*, Avebury, Aldershot.

Katsikides, S. (1994), *Informatics, Organization and Society*, Oldenbourg Verlag, Wien-München.

Layton E, (1971), *The revolt of Engineers*, Case Western University Press, Cleveland.

Linde, H, (1972), *Sachdominanz in Sozialstrukturen*, Tuebingen.

Linde, H, (1982), *Soziale Implikationen technischer Geraete, ihrer Entstehung und Verwendung*, in R. Jokisch (Hg.) Techniksoziologie, Frankfurt am Main.

Loewith, K, (1964), *Das Verhängnis des Fortschritts* in H.Kuhn, F.Wiedmann (Hg.) Die Philosophie und die Frage nach dem Fortschritt. Verhandlungen des Siebten Deutschen Kongresses fuer Soziologie, München.

Meiksins P. (1982), *Science in the Labor Process: Engineers as Workers*, in Charles Derber (ed.) Professionals as Workers: Mental Labor in Advanced Capitalism, G.K.Hall and Co., Boston.

Noble David, 1977, *America by Design*. Oxford University Press, Oxford.

Markuse, H. (1967), *Der eindimensionale Mensch*, Neuwied-Berlin.

Markuse, H.(1941), *Some Social Implications of Modern Technology*, in Zeitschrift fuer Sozialforschung, Jg.9, Nr.3/41.

McNeill M. (1989), *Research Methods*, 2. edition, Routledge.

Mowery D and Rosenberg N, (1979), *The Influence of Market Demand upon Innovation*, Research Policy, Vol.8 No.2.

Mumford, Lewis. (1977), *Mythos der Maschine* Fischer-alternativ-Verlag, Frankfurt a.m., p. 493.

Ogburn, W, (1969), *Kultur und sozialer Wandel*, Neuwied/Berlin.

Rammert, W, (1983), *Soziale Dynamik der technischen Entwicklung*, Opladen.

Rammert, W, (1991), *Entstehung und Entwicklung der Technik*, Stand der Forschung zur Technikgenese in Deutschland. WZB, Berlin.

Rammert, W, (1992), *Wer oder was steuert den technischen Fortschritt?* Soziale Welt, 43 (1), p. 83.

Rammert. W, (1993), *Technik aus soziologischer Perspektive*. Opladen.

Salzman, H. (1989), *Computer Aided Design: Limitations in Automatic Design and Drafting*, in: IEEE Trans. on Engineering Management, Vol. 36, No. 4.

Salzman, H., Rosenthal, S., (1994), *Software by Design*. Oxford University Press, New York.

Schmookler, J, (1966), *Inventions and Economic Growth*, Mass. Harvard University Press, Cambridge.

Széll G., (1994), *Technology, Production, Consumption and the Environment*, International Social Science Journal, June 1994 Vol. 140, Blackwell, Oxford.

Tschiedel, R, (1989), *Sozialverträgliche Technikentwicklung*, Oblanden.

Tschiedel, R, (1990), *Die technische Konstruktion der gesellschaftlichen Wirklichkeit*, Muenchen.Alemann et al. 1992.

Szèll G. (1994). *Sociology: State of the Art II*, Political, economic and socio-demographic dimensions, in: International Social Science Journal, Blackwell Publishers, Vol. 140, xlvi, No. 2, June 1994.

Ulrich, O, (1979), *Technik und Herrschaft*, Frankfurt am Main.

Weingart, P, (1989), *Technik als sozialer Prozess*, Frankfurt am Main.

Whalley, P. (1986), *The Social Production of Technical Work: The Case of British Engineers*. London: Macmillan, Albany: State University Press

Zussmann, R. (1985), *Mechanics of the Middle Class: Work and Politics among American Engineers*, Berkeley: University of California Press.

8 Critical Issues for the Domains of an Information Society

Part One

Though this chapter will focus on different issues of information society, from a critical point of view, it is essential to recall some thoughts from chapter two and chapter four in order to outline theory and empirical evidence.

Chapter two has attempted to show that there is a variety of theoretical issues which can be directed to mainstream sociology of technology. One common understanding which derives from the research in the field reveals that most sociological studies on technology use the comparative method, and the remaining surveys apply to the field of technology assessment. We have argued that technology reflects the synergy of power and societal processes, and these must be analyzed under the foci of sociology of science or even of the emerging sociology of information. While sociology of information should address a variety of theoretical perspectives that can be directed towards the social phenomenon of information, they alone do not give sufficient insight into the nature of information either as an object of disciplinary discourse or as an object of nature (Balnaves 1993:108).

The emerging approach is that an entirely new concept is required and that there is a vital need for improved analysis with respect to the assessment of technological issues. It can be argued that theoretical considerations have to be linked with practical methodology in order to evaluate technological and societal issues, because different sets of complexities exist between the cultural and the operational aspects of the functional role of technology. However the issue here is more complex, and the argument can be summarized as follows. The first problem relates to methodology, where it is clear that a global approach, whether theoretical or empirical, reaches its limits very quickly. The second problem is a more general issue that refers to all the social sciences: a common direction to resolve common social phenomena is lacking. Thirdly, it can be argued that a new approach is needed, which would focus on a detailed evaluation and provide a synthesis of all the intervening variables involved in the technological discussion. One example of such an approach is the ARS model (Katsikides 1994). Finally, technological developments, like other social, economic, and technical approaches, are not socially neutral, and in the end they deal with different traditions (European, US, Scandinavian, Japanese, etc.). As such they accumulate social processes and reflect them, or, as Thomas Kuhn (1970) put it 'a failure to assimilate fully new conditions and technology will strain the existing structures' of society. Existing structures means the certain sociopolitical and economic span will continue in a sense to construct the system, where modernization has a long way to go. If we can see technology as a social phenomenon, by terms of sociological analysis and as such determined, then is socially constructed, this however, is definitively a case of socialization and thereby could be socially 'taught and learned'. Chapter two ends with questions that could make people rethink their choices concerning technological change and problem solving.

Problem solving is a new parameter in the technological discussion and creates possibilities for redesigning the workplace, and offer solutions, which arise through the adoption of information systems and their impact on labor. A further key aspect in this work are the case studies in French hospitals. (Katsikides/Schneider 1994). In chapter four, the problem arises in different hospitals and various models were proposed and finally adopted. The hospital sites analyzed in the survey were selected from the public and private sectors. For the most part, they were attached to a university teaching hospital center. All institutions had made or were planning to make investments in computing equipment, that besides handling financial management tasks, could also provide support for the organization of a patient's hospitalization from the perspective of a medical-technician or the clinician. The study focussed on the information system development and acquisition policies applied, affecting the main characteristics of the target system as well as the user involvement. Four hospital institutions are discussed. The first case is a pediatric city clinic which runs with an integrated hospital information system that was developed externally by a major computer manufacturer. The second case is a private institution for the treatment of cancer diseases, equipped with an internally developed system. The third case is the surgical department of a cardiological hospital operating with a highly complex local system combining the time-critical demands of the ICU setting with the general care requirements of pre- and postoperative treatment in the ward. The fourth case is a regional hospital center maintaining a loosely coupled decentralized architecture of local networks.

Part Two

Hulin and Roznowski (1985:47) define in their work *Organizational*

Technologies: Effects on Organizations Characteristics and Individuals Responses and observe technology as the 'physical' combined with the 'intellectual' of knowledge processes by which materials in some form are transformed into outputs used by another organization or subsystem within the same organization. Law (1987:115) for instance, saw technology as a family of methods for associating and channeling other entities and forces, both human and non-human. It is a method, one method, for the conduct of heterogeneous engineering, for the construction of a relatively stable system of related bits and pieces with emergent properties in a hostile or indifferent environment. Berniker (1987:10) claimed that technology refers to a body of knowledge about the means by which we work on the world, our arts and our methods. Essentially, it is knowledge about the cause and effect relations of our actions. ... Technology is knowledge that can be studied, codified, and taught to others Weick (1990:2) for instance, defines technology as equivogue that is something that admits of several possible or plausible interpretations and therefore can be esoteric, subject to misunderstandings, uncertain, complex and recondite. He provides a context that illustrates the strengths and weaknesses of prevailing thought about technology. Weick provides three definitions of technology. After reviewing those, we begin our analysis of different social scientific approaches towards technology.The first definition, however, addresses in Weicks understanding the explicit mention which is made of raw materials and a transformation process, items that are often implicit in other definitions. Also novel to this definition is the mention that output might be used within the same organization. It is possible, Weick, stated, that diverse technologies within the same organization could exist. Finally, this definition is noteworthy because of its emphasis on processes rather that on static knowledge, skills and equipment. By equating technology with process, Hulin and Roznowski alert us to the importance of changes over time and sequence (Weick, ibid).

The German philosopher Heinz Hülsmann, (1916-1993) in his well discussed book *Die technologische Formation* (1985), which has not yet been translated into English, pointed out, when dealing with the idea of technology its link to formation:

> Technik ist nicht nur Können und Kunst, Fertigkeit und Wissen, sondern eben auch List und damit Gewalt und Herrschaft,wie dieses im Verhältnis der Form zur Materie sichtbar wird und wirksam ist.

Free translation:

> Technology is not only ability and culture, proficiency and knowledge, but

also slyness and therefore violence and domination as their relation can be seen and gain efficiency to form and subject.

Hülsmann stated clearly the direction which sociological work takes when examining technology and its shaping. Although his book was published in the mid 1980s underlines the distinction, in my eyes between the European and the American context, which we will try to distinguish in our discussion. Salzman (1994) noted in his work, that the Scandinavian approaches, for instance could provide a model for ways to reorient design but also showed that their particular approach reflects a particular industrial culture (socialization of workers, engineers, etc.) shaped by, among other factors, a longstanding craft tradition, a workplace environment of 90 percent unionization, and requirements for labor union participation in many basic decisions that would be considered management prerogatives in the United States. A further point which has been made by D. Jahn[1] (1994), indicates the differences between the German and the Swedish trade unions; his findings demonstrate the differences in both countries. In Sweden, in contrast to the assumption expressed in the 1970s, technological progress has not been challenged to a larger degree. As a consequence there was no increasing tension between labor and capital or within the trade union movement.

A further definition of technology which was made by Law (1987) and analyzed by Weick, implies that the design and operation of technology do share some of those qualities but they do not exhaust the character of the process when it unfolds in politicized organizations, and the above mentioned definition allows us to describe technology in a way more compatible with this quality of organizations, (Weick 1990:4). The third and last definition by Berniker, states that every technical system embodies a technology. It derives from a large body of knowledge which provides the basis for design decisions. Weick (ibid) in his own analysis argued that the first definition forces actually to re-examine our knowledge of cause effect relations in human actions and the choice of a different combination of machines, equipment and methods to produce the outcomes for which new technologies are instrumental. He continues, pointing out Berniker's argument and stating that technology follows rather than precedes a technical system, and, furthermore, that technology is both an a posteriori product of lessons learned while implementing a specific technical system and an a priori source of options that can be realized in a specific technical system. All of these analyses respond only partly, to the different debates and approaches when mapping technological and sociological knowledge. Furthermore, as Weick illustrates, other works by Scott (1987), Hancock, Macy and Peterson (1983) and Perrow (1986), provide helpful summaries when definitions of technology have been translated into survey items intended to capture variations in skills, equipment, and technique. Weick on this issue concludes that new technologies introduce a set of issues that organizational theorists have yet to grapple with. Unless they do, the power of technology as a predictor of organizational functioning will diminish.

As now organizations are concerned with the introduction and useful usage of new technology, it seems that old industries or other enterprises can neither be compatible as previously Law (1987) stated, nor follow a certain new form of organizational change, which is influenced by external parameters. Emerging technologies force both the organizational structure and external relations to become more efficient and proceed to optimization. R. Reddy(1990:249) analysing the new forms of organization concerning the US car manufacturing industry, which have to be more efficient, concludes that:

> if the United States continues to take four to five years to introduce a new car while Japan can do it in half the time, then obviously Japan will continue to increase its market share.

Let us recall the three logo model (LM), which have been discussed in Chapter Two and their application. A further point which can be made is the diversification between the European and the US technological concepts. The European concept is more state dominated (EU control) and several options and regulations, which have been approved by the Commission, are for instance, now not allowing subventions from the fifteen states and their regional local governments to industrial and other business conglomerations. The policies of these conglomerates against the EU regulations can be viewed each year in a list of firms which have to pay a penalty for ignoring these rules, or simply for violating them.

The first perspective of the LM concerns the organizational structure, which can be found in every enterprise or organization. The second, extrovertial, perspective shows the sphere of action of the organization on the outside world. The third, introversial, perspective covers the sphere of action in the inner life of the organization, where all processes within the organization are functioning. As we have seen, the structural change in organizations covers more and more enterprises. Public and private organizations are building their organizational structure on a common criterion. As a functional connection of all management and administrative starting points in the inner life of organizations apart from their size, joint criteria can be observed. The observation of this coherence will be shown in a shaping LM where the minimum of a organizational structure will be taken as a basis for new technology as the next logical path of other forms. The results of the various adaptations and applications of information technology, i.e., in production such as CAD, CIM, CAM etc. lead to the argument that (1) automation and rationalization effects were the first issue and (2) the final result was the flexible oriented production. All these systems require a new method of administration. A new theme of organization which includes fields and approaches such as data transmission, telecommunication, innovation of production, rationalization of working operation, of employers etc. is emerging. R. Kling[2] (1994) states on this particular point:

> many organizations are adopting computing equipment much more rapidly than they understand how to organize positive forms of social life around it. However, some fervent advocates of computerization portray the actual pace of computerization in schools, offices, factories, and homes as slower than they wish. These 'computer revolutionaries' argue that many key institutions - such as schools, businesses, family life, public agencies - can be progressively reformed through the appropriate application of computer-based systems.

It should be now clear that changes in organizations might deliver the reason for other compatible tasks, which are included in the planning and might be adopted and implemented later. Hartmann (1986:180) makes the point that the crisis of the administrative work forms the compulsion of continuing production. The second point concerns the administrative operations which must go faster when necessary. The third point is the assistance to the administrative operation taking in account the flexibility of the enterprise; that means a faster collection and distribution of concrete transformations of information and data. It is not surprising therefore that the installed system which creates new organizational structures, was established as an instrument which operates independently of political and social decisions. That implies that the organization is not in the position to control the system anymore. The decisional parameters lie outside their action fields.

The second concept after the structural perspectives of the LM is to be seen on the external action radius of the organization. The organization analyses the relationships of the enterprises with the world outside. Examples are mother and daughter companies, the relations with the state, the law, trade unions and other interested organizations and last but not least the customer. The compatibility of an enterprise is obvious through the synergies in the level of employment, on the concepts management and finally sales and marketing area.

The third observation should be the action sphere which is to be found in the internal structure of the enterprise or organization. At this point we analyze trade unions, content of work, working time, collective bargaining, agreements, security, creation of work, economics, etc. The question which now arises is whether the US and the European technological concepts could in same way be compared and measured, according to the guidelines of European Union for the creation of a United Europe?

Part Three

Further analysis has been undertaken by Egger E. (1995), in her book *CSCW: The bargaining aspect*. From the informatics' designing point of view, Egger has carried out a set of surveys which contribute to Computer Supported Cooperative Work (CSCW), that means to the development of technical systems which at the same time support the work of groups. The innovative idea here is that CSCW integrates tools for working teams. That means it creates new measurements for work based on also on the social environment. Through CSCW, which could become a vital issue for sociologists as Egger (1996:13) pointed out and has examined not only obvious technological systemic aspects but also the role of sociology. CSCW claims to support the working processes of groups. Therefore it is an interesting issue to analyze how people work together and which aspects of group dynamics have to be considered when designing technical systems, especially those focusing on planning processes. Therefore the contribution of sociology in defining design guidelines for CSCW systems has to cover working groups and group processes, bargaining situations and the role of time in groups. Controlling and monitoring were the first goals of the first generation of computer technologies, which have been implemented in production processes. Sydow (1985) used the term 'joint optimization' describing the idea of combining the advantages of technical systems and human qualifications. Later the new idea which was to match principles stemming from the movement of industrial democracy, industrial psychology and usability of technical systems, was carried out by Kling (1984), Olson (1983) and Shaiken (1985). Bannon and Schmidt (1991), trying to define the work process, stated that, 'cooperative work is constituted by the work processes that are related as to content, that is, processes pertaining to the production of a particular product or service'. Holand and Danielson (1989) suggest three kinds of perspectives of cooperation(see Egger 1996:27):

(1)   Cooperation as a strategy: cooperation is based on solving disagreements and conflicts where the participants state their position very clearly. In the necessary processes of solving conflicts some participants try to persuade the others or try to push their interests by

building coalitions.

(2)   Cooperation as coordination: cooperation is described as a method for a group to solve some joint problem or perform a common task. The process of cooperation is based on a sharing- among all participants involved- of the responsibility of reaching the goal.

(3)   Cooperation as reflection and creativity: cooperation is interpreted as a group process where the partners are encouraged to contemplate and reflect on the matters being discussed. All participants are described as being potentially of equal interest for the observer.

Bannon and Schmidt (1991) in their work *CSCW: Four Characters in Search of a Context*, define CSCW as work by multiple active subjects sharing a common object, supported by information technology. A common object of work is to draw a need to a 'shared goal' which is criticized as being too restrictive and to 'shared material' which is criticized as being too loose. (see Egger:28).

Conclusion

D. Edge[3] (1995:15) has pointed out two interesting approaches concerning the social shaping of technology. The first deals with the history and sociology of science and goes back to Pinch and Bijker (1984). The content of this conception is to study the development of technological fields, whereby it is essential to identify points of contingency or interpretative flexibility, where at the time ambiguities are present. When such 'branch' points have been identified the researcher then seeks to explain why one interpretation rather than another succeeded or why one way of designing an artefact triumphed. The second approach works 'in' from the context. Here the starting point is the particular social context within which technical change takes place. The focus is on everything which contribute to shape technology. P. Senker (Senker 1995), makes an interesting point concerning IT and the role of the developing countries in perspective.

H. Mackay[4] (1995:41) in his contribution *Theorising the IT/Society Relationship* gives an overview of the sociology of technology, pointing out that sociologists have until recently tended to avoid technology, but than this began to change significantly in the late 1980s with the growth and development of both (physical) IT and the (social) debate surrounding it. He is on the right path, when he stated that:

Sociologists of technology are concerned with explaining how social processes, actions and structures relate to technology; and in this are concerned with developing critiques of notions of technological determinism. The theories and concepts which have been developed are increasingly recognized as of value to technologists, notably in the area of information systems design.

If technological determinism is the concept that technology is autonomous, then symptomatic technology as stated by R. Williams (1974) explains the inverse, namely, that technology is a symptom of social change. Mackay further, (1995:41), explaining this point, stated that 'according to this model, it is quite clearly society which is in the driving seat of history: given a strong social demand then a suitable technology will be found'.

Although from lab invention to wide market consumption it might sometimes takes up to seventeen years, (Profil 1997:19) it should be noted here, that other socio-political and economic factors play a vital role in determining whether a certain technology is adopted or not. Braveman (1984) arguing about technology, also makes the point that it can not be focussed only on individual inventions. Furthermore, it is needed to examine how broader socio-economic forces affect the nature of technological problems and solutions (see also Mackay 1995:43).

Notes

1   Jahn, D. (1994), *The Challenge of Technological Progress* in Modern Societies. In: Katsikides, S. et al., (1994), *Patterns of Social and Technological Change in Europe*, Avebury, Aldershot. Jahn's point: When looking at Germany, we see that in the dimension of productionist/alternative world views the gap between capital and labour has widened. Since the compromise on societal development oriented towards technological progress and economic growth is a basic principle for industrial society, this gap may lead to the conclusion that there are some hints for an undermining of the societal consensus. However, this cleavage does not run very nicely between labour and capital, but instead it shows that there is a cleavage between parts of labour, on the one hand, and capital and other parts of labour, on the other. This conclusion even leads to the fact that we may expect a rising cleavage within the labour movement. This cleavage runs along the lines of 'radical' versus 'moderate' factions in the labour movement. The early 1990s have become challenging times for the ideas that criticise productionism as economic security and nationality

has reasserted itself throughout Europe. The revolutionary processes in Eastern Europe and the unification of the Federal Republic of Germany may have led to a decrease in the importance of the cleavage analysed in this article. Particularly the standpoints of the proponents of technological progress have been supported by this development and material aspects gained in importance. At the moment, it is difficult to predict whether or not this trend will continue or if the opposition against the dominant industrial culture will become alive again. However, the form and content of such protests can change significantly in that there is a movement from a protest that emphasises the irrationality of social development and fragmentation of culture by stressing egalitarian values, towards a protest that hopes to interrupt the process of alienation by marginalization of underprivileged groups and stressing national values'.

2   Kling, R. (1994), *Usability versus Computability: Social Analyses by Computer Scientists*. In: Katsikides, S. (ed.), Informatics, Organization and Society, Oldenbourg Verlag, Wien- München, Kling states further (1994: 112-113) as Perspective four: Professional Responsibility to Society is Essential, in the above mentioned paper, the following: Some computer specialists are specially concerned that computing technologies should be 'sound products'. According to this view the computing professions should be responsible to their clients by delivering practical systems which are usable, reliable, and safe. The main foci of attention have been to identify computer systems which can be major threats to physical safety or civil life and to identify discrete solutions to reduce these risks. Some forms of computer technologies can be harmful because of unreliable software (e.g., life-critical information systems; election counting systems; social security payments; fly-by-wire aircraft; military command and control systems.). Other kinds of computer based systems threaten to diminish personal privacy. Both kinds of threats have been the subject of a specialized topical literature, the concern of organizations like Computer Professionals for Social Responsibility and special forums, like the ACM sponsored 'Risks' computer bulletin board. In this perspective the primary reforms will come through improved software quality and certain changes in organizational practices (e.g., privacy protections; administrative guidelines to insure safe software and data handling practices).

3   Edge, D. (1995), *The Social Shaping of Technology*. In: N.Heap, et al., Information Technology and Society, A Reader. Sage. For Britain, Edge notes (1995:.28). 'How, then, might Britainãs performance be

improved in this respect? How can we increase attention to the entry into use, and more widespread adoption, of new technologies, and to strengthen the influence of the exigencies of these processes on the generation of new 'basic' knowledge, and the development and design of new products? Exhortations to 'so better', or to be 'more like the Japanese', are simply not going to be enough. For, typically, the different types of knowledge involved in the different phases of the product-cycle model are possessed by defferent types of people, and there is hierarchy of prestige that rises as we move from right to left across our diagram. 'Basic research' is more prestigious than the mundane -but commercially crucial- tasks of implementation, marketing, distribution,maintenance and repair. All these involve technological knowledge, but it is knowledge our culture typically devalues'.

4   See also: Senker, P. (1995), *Technological Change and the Future of Work*. In: Heap N et al. (1995), Information Technology and Society, A Reader, Sage. p. 146. Senker makes an interesting note on the role of IT, its perspectives and the role of the developing countries. He notes: Japan shows every sign of maintaining its technological lead in IT and advanced materials, with the USA following, and, perhaps, a growing gap between Japan and the USA on the one hand and Europe on the other. New Technologies are likely to be expoloited mainly to provide better paid and more interesting jobs in the countries which dominate their initial development and exploitation. The main benefits in terms of production and development are likely to accrue to the more advanced countries-in the next few decades a category which may include the advanced regions of some newcomers such as South Korea and Brazil. Nevertheless, many developing countries seem likely to fare relatively badly. Bio-technology, for instance, is likely to incrase rapidly in significance in the early years of the twenty-first century. It is unlikely that developing countries will benefit disproportionately in terms of employment: indeed, such trends as it is possilble to discern indicate that advanced countries may well benefit at the expense of developing countries.

References

Balnaves, M. (1993), *The Sociology of Information*, ANZJS Vol. 29, No.1, March 1993, p.108:
Bannon L. and Schmidt K.(1991), CSCW: *Four Characters in Search of a*

Context, in: Bowers J.M. and Benford S.D. (eds.), Studies in Computer Supported Cooperative Work, North Holland, Elsevier, Amsterdam.

Berniker, E. (1987), *Understanding Technical Systems*, Paper presented at the Symposium on Management Training Programms: Implications of New Technologies, Geneva, Switzerland, Nov.1987, p.10, also cited by Karl E.Weick, Technology as Equivoque: Sensemaking in New Technologies. In Goodman S.P. et al. (eds.), Technology and Organizations, Jossey-Bass Publishers, San Francisco, Oxford, 1990, p.3.

Egger, E. (1996), CSCW: *The Bargaining Aspect*. Vienna/Frankfurt/New York: Peter Lang.

Giddens, A. (1976), *New Rules of Sociological Method,* London.

Giddens, A. (1982), *Profiles and Critiques in Social Theory*; Berkeley, University of California Press.

Giddens, A. (1984), *The Constitution of Society. Outline of the Theory of Structuration*, Cambridge.

Hartmann, M.(1986), *Strategien und Resultate der Verwaltungs-rationaliserung*. In: Journal für Sozialforschung, Heft 2, p. 180.

Hancock, W.M., Macy, B.A., Peterson, S. (1983), *Assessment of Technologies and Their Utilization*. In S.E. Seachore, E.E. Lawler III, P.H. Mirvis and C. Cammann (eds.), Assessing Organizational Change. New York: Wiley.

Heap, N., Thomas, R., Einon, G., Mason, R., & Mackay, H., (1995), *Information technology and society*, Newbury Park & London, Sage.

Holand U., Danielson T. (1989), *The Psychology of Cooperation-Consequences Descriptions*, TF-Report 58/89, Oslo.

Hulin, C.L. and Roznowski, M.(1985), *Organizational Technologies:Effects on Organizations' Characteristics and Individuals Responses*. In: L.L. Cummins/B.M. Staw, (eds.), Research in Organizational Behavior. Vol.7. Greenwich, Conn. JAI Press, p.47.

Olson, M. (1983), *Remote Office Work: Changing Work Patterns in Space and Time. Communication of the ACM*, 26(3) March, pp. 182.

Katsikides, S. (1994), *Interests in the Transformation of Organizations*. In: Katsikides, S. (ed.), Informatics, Organization and Society, Oldenbourg Verlag, Wien- München, p.47.

Kling, R. (1984), *Assimilating Social Values in Computer-based Technologies*, Telecommunications Policy (June), pp.127.

Kuhn, Th. (1982), *The Structure of Scientific Revolutions*, Chicago University Press.

Law, J. (1987), *Technology and Heterogeneous Engineering: The Case of Portuguese Expansion*. In W.E.Bijker, T.P.Hughes, and T.J.Pinch (eds), The Social Construction of Technological Systems. Cambridge, Mass.:

MIT Press, p.115.

McLellan, ed. (1988) *Marxism: Essential Writings*; Oxford University Press, Oxford.

Perrow, C. (1986), *Complex Organizations*. 3rd ed. Random House, New York.

Profil. (1997), *Technische Revolution*, Nr. 27, 30 Juni1997, 28 Jg. e19, Wien, Austria.

Reddy, R. (1990), *A Technological Perspective on New Forms of Organizations*. In: P.S. Goodman et al. (1990), Technology and Organizations, Jossey-Bass, San Francisco, Oxford, p. 249.

Scott W.R., (1987), *Organizations: Rational, National and Open Systems*. Englewood Cliffs, Prentice-Hall, N.J.

Shaiken, H., (1985), *The Automated Factory: Vision and Reality*, Technology Review 1/85, Washington.

Williams R., (1974), *Television Technology and Cultural Form*, Fontana, London.

9 An Interpretation of Sociology in the Information Society

This session discusses a general view of sociological work, drawing on recent approaches on the 'philosophy' of information and communication systems and its theoretical links to social theory. It reviews different arguments about information technology and reexamines contemporary ideas about social theory. This chapter concludes, first, that the post industrial society derives from different scientific and societal traditions and as such accumulates and reflects power and in the end creates a new political and societal condition. Secondly, at the same time it should be noted that information systems are much more a condition of human action instead of a tradition, which serve as the pathmaker for modernity. Finally, rethinking sociological work as it is, and developing concepts as model simulation are the next steps in sociology, as suggested.

Introduction

What is the common sociological understanding explaining the term social theory? The answer is methodology and more specifically the logical and philosophical questions that study humans and their social world. Taking as an example a science fiction story from *Star Trek*, let me symbolically compare reality and how real is virtuality and what science fiction mean for the social world.

> *Odo*, beyond the wormhole in the Gamma Quadrant, sitting in a small shuttle ship and traveling back to DS9 base, speaking to the board computer, take us *home*..............eh...what's that!
>
> (Star Trek, Deep Space Nine).

Odo the Constable, responsible for the security of the *Deep Space Nine*, refers to the word 'home' and indicates that in this place of the universe, home is not the home term as we may now know, but exactly the contrary. Home is the place where we live. This could also be compared with postmodern thinking and eventually this is the reason why we refer to a non-real story which definitely is enriched with human values and has greater importance on the people in our epoch. What is then more real than this? Postmodernist should think? An impact on society now from a story, even when it comes from the future? or real stories from now, without any impact on nobody, not today and not tomorrow. Finally everything depends on interpretation. Discussions of similar kind have started in the begin of our century and created real directions or schools of thought in all kind of arts and particular in science.

If somebody follows discussions on the level of Alan Sokal and Jean Bricmont in their widely discussed book *Impostures Intelectuelles,* and its controversy with Jacques Derrida, then is easy to understand the interplay. Derrida has written that Sokals' text which was sent for publication to the American Journal *Social Text,* and was accepted there, was a hoax (the Sokal's hoax). Sokal and Bricmont have accused, French intellectuals as Barthe, Derrida, Althusser, Foucault, among others, that their so called post-modern philosophy is valueless and describe more or less exactly what is sees, and this by their own interpretation rules. (They argue that postmodernism remains beyond scientific methodology and historicism). Eventually, what postmodernist's describe is not the scientific truth, as they indicated. Foucault, for instance, takes care to emphasize that by 'truth' he does not mean 'the ensemble of truths which are to be discovered and accepted'. By 'truth', he means the ensemble of rules according to which the true and false are separated and specific effects of power attached to the true. The struggles around truthare not 'on behalf' on the truth, but about the status of truth and the economic and political role it plays (J. Whitehead[1] 1994:149).

This discussion has, however, initiated more discussions in the most European countries, where intellectuals try to argue based on the above mentioned dispute on the meaning of postmodernism for there *national* and therefore for the entire science.

These schools of thought, as mentioned above, developed a certain way of writing, which definitely brakes up with the traditional writing, or text interpretation. Jacques Derridas' critique on the philosophical tradition can by other people only shared and overtaken. Because is a certain ideological and political position as Lange (1975, 248) points out in his book *About Contemporary French Literature.* Furthermore, there are more questions as answers, because either you accept the term, *critique of metaphysic* or you leave it. Empirical materials are here not allowed. On the other hand, Derridas work with literature, creates so far not a model where its usage can be applied in other sciences and disciplines, which they deal with text interpretation.

To return back to our Star Trek example, it must be indicated here that scientific praxis has its limits due to virtuality, when this deals primarily with imagination and metaphysics. In order to proceed further, however, with the real target of this work, I have to disappoint the *trekkies* by saying that in fact we cannot analyze imagination, at the same time when we try to analyze social theory, which obviously reflects reality. The construction of social theory itself, namely, methodology, analysis of modernity and social critique creates enough tools for referring to movements within the arts and why not, in science. To the second discussion point, it must be said that its not possible here to go deeper discussing modernity. However, the term modernity as Abercrombie et al.[2] stated refers to a movement within the arts in Western societies between about 1880 and 1950, as represented by figures such as *Picasso* in painting, *Eliot* in poetry, *Joyse* in literature, *Stravinsky* in music and *Bauhaus* in architecture. Above all emphasized novelty, although by the middle of the twentieth century it had almost become the orthodoxy, as they characteristically said. They argue that it has been superseded by postmodernism (Abercrombie et al., 269). On the contrary, postmodernity argues for a condition which contemporary advanced industrial societies are alleged to have reached. A very large number of features are said to characterize postmodernity and they may be placed into four groups-social, cultural, economic and political (ibid: 326).

The paper is organized as follows. The first section deals with definitions and sets prerequisites for the second section, where the main target is to analyze technology as a social system.

Part One: Social and Cultural

H.Salzman and S. Rosenthal, (1994) have focused as social scientists specifically on the design of workplace technology and showed how software design and usage leads to essential tasks of engineering which involves social values. These social values reflect the economic and political structures of organizations and the values provide the background assumptions shaping peoples' perspectives of their world of work. Social critics were also addressed at an early stage by Lewis Mumford (1934) who saw it as a problem of technological society and autonomous technology. Jacques Ellul (1964) warned of technological dominance of human life with ensuing impoverishment of the human spirit. From a critical point of view we can mentioned David Noble (1984) who observes that:

> although it has belatedly become fashionable among social analysts to acknowledge that technology is socially determined, there is very little concrete historical analysis that describes precisely how.

It can be said for Noble's pioneering work that it has developed a growing interest in and body of research on the social shaping of workplace technology. Other useful works in this direction were e.g. Hochgerner (1986), Bijker, Hughes and Pinch (1987), Broedner (1990) or Corbett, Rasmussen and Rauner (1991), R. Kling (1992), Winner (1977) and Rammert (1992). Nevertheless, traditional thinking in industrial relations focussed more or less until the end of the 1970s on the sociology of work and curried out remarkable studies which have analyzed the relationship between employment and the institutions which are associated. It embraces the relations between workers, work groups, worker organizations and managers, companies and employer organizations. The study of industrial relations is an interdisciplinary enterprise, drawing heavily on industrial sociology, labor economics and trade-union history, and to a lesser extent on psychology and political science (Abercrombie et al., 213).

As Anthony Giddens defined, *social theory* is not a term which has any precision, but it is a very useful one, as he characteristically indicates. Further, he points out that social theory involves the analysis of issues which spill over into philosophy, but it is not primarily a philosophical endeavor. The social sciences are lost if they are not directly related to philosophical problems by those who practise them. To demand that social scientists be alive to philosophical issues is not the same as driving social science into the arms of those who might claim that it is inherently speculative rather than the empirical. Social theory has the task of providing conceptions of the nature of human social activity and of the human agent which can be placed in the service of empirical work. The main concern of social theory is the same as that of the social sciences in general: the illumination of concrete processes of social life. (Anthony Giddens, *The Constitution of Society*, xvii, Polity Press). Recently Holmwood[3] (1995), in an article in BJS and while dealing with feminism and epistemology outlines the argument that:

> The challenge for social theory is to re-construct its explanatory categories, rather than to de-construct the explanatory undertaking. Postmodern theory is a capitulation in the face of our problems, rather than any solution of them. It does, indeed, embed contradiction in its theory of knowledge, as Hawkesworth suggests.

As we have pointed out in previous works (Katsikides 1997), an effort has to be made to understand societal transitions and methodological means, or, as Talcott Parsons advocated, using sociology to study the relationship of an individual's experience to society and history, the starting point for the sociology of technology must be through science. For industrial societies which recognize 'growth' as a constitutive necessity, constant change could even guarantee preservation. Such a societal development can only be maintained continuously if it is able to remodel itself by adopting to constant change in a controlled way. This regulating mechanism which controls human behavior according to the specific needs of a certain given societal development, is termed a formative principle. It organizes existence and change in social behavior over several historical eras without itself being restricted to the respective form of that time. On the other hand, objectivism deals with human behavior when this is standardized, or 'functioning'. Finally, the study of what has come to be termed as the sociology of technology does in fact incorporate elements of sociological methods, since they can illustrate social behavior in a regulated societal system where technology and formal foundations create the context and the perspective. In the above mentioned work (Katsikides 1997) the aim was to show that there is a variety of theoretical issues which can be directed to mainstream sociology of technology. One common understanding which derives from the research in the field reveals that most sociological studies on technology use the comparative method, and the remaining apply to the field of technology assessment. We have argued that technology reflects the synergy of power and societal processes, and these must be analyzed under the foci of sociology of science or even of the emerging sociology of information. While sociology of information should address a variety of theoretical perspectives that can be directed towards the social phenomenon of information, they alone do not give sufficient insight into the nature of information either as an object of disciplinary discourse or as an object of nature (Balnaves[4] 1993:108).The emerging approach is that an entirely new concept is required and that there is a vital need for improved analysis with respect to the assessment of technological issues. It can be argued that theoretical considerations have to be linked with practical methodology in order to evaluate technological and societal approaches, because different sets of complexities exist between the cultural and the operational aspects of the functional role of technology. However the issue here is more complex, and the argument can be summarized as follows. The first problem relates to methodology, where it is clear that a global approach, whether theoretical or empirical, reaches its limits very quickly. The second problem is a more general issue that refers to all the social sciences: a common direction to resolve common social phenomenon is lacking. Thirdly, it can be argued that a new approach is needed, which would focus on a detailed evaluation and provide a synthesis of all the intervening variables involved in the technological discussion. One example of such an approach is the ARS model (Katsikides 1994). Finally, technological developments, like other social, economic, and technical approaches, are not socially neutral, and in the end they deal with different traditions (European, US, Scandinavian, Japanese, etc.). As such they accumulate social processes and reflect them, or, as Thomas Kuhn (1970) put it 'a failure to assimilate fully new conditions and technology will strain the existing structures' of society.

Moving now forward in order to analyze the position and values of technical workers, several ideas and notions can be made. S. Crawford, for example, in his work *Technical Workers in an Advanced Society*, published by Cambridge University Press in 1988, has pointed out that only two empirical studies compare the position and values of technical workers in old and new industry. Zussman's (1985) book on American engineers and Whalley's (1986) volume on British engineers. Crawford introduces in his above mentioned book the study of French technical workers both complements and builds on their important work. He positioned that the findings of an investigation into the work, careers, and ideologies of French engineers and managers employed in two industrial settings, a traditional metal-working firma and an advanced telecommunications firm. What makes actually this notion so interesting is the passage from the theoretical considerations over to the real life on technological systems. The question, who drives whom either technology or society, and *vice versa*, seems that it has reached a dead end. In this chapter, engineers are the object of investigation.   Several empirical studies as Crawford (1988) stated, have tried to describe the largely unsuccessful struggles of American mechanical, electrical and other engineers to develop codes of ethics, unify, and form an association powerful enough to effectively represent their interests of employers and the state. It must be taken into consideration that other aspects of the advanced society and its industrial systems also play an important role in this issue. Trade Unions, for instance, education and training, collective agreements, and market performance, play a vital role on technological decisions. Contemporary status and identity of engineers, was the aim of the work of Durand (1972). Some of these works were, Calhoun (1960) in his book *The American Civil Engineer* or Layton (1971) in his work *The revolt of Engineers,* Noble (1977) in his pioneering book *America by Design* or Meiksings (1982) with his work *Autonomy and the Engineer*. Further studies in France as Grelon (1986) *La Modèle de l'école d'ingénieurs comme formation à la technologie et comme insertion dans la société* on the history of French engineers, analyse the historical structures. The outcomes of all these works were characteristic of their epoch, the idea, which was highlighted was if we can't change the industry or the capital behind it then the engineers. This conception, however, finally failed. New industries, are new technology oriented structures, our aim is to look at several industrial enterprices, where information systems, electronics and telecommunications employ 'technicians', engineers and technical workers.

Salzman (1995) pointed out that the main view of engineering is that it is applied science, that is, it is the application of scientifically and objectively determined principles. The 'scientific view' of technology is that advances in knowledge are largely independent of subjective influences. Thus, technology reflects engineers' calculations of the most economic and efficient designs to utilize that knowledge. This is the dominant view of engineering as expressed, for example, in the U.S. by the accreditation Board for Engineering and Technology, definition of engineering as 'the profession in which a knowledge of the mathematical and natural sciences gained by study, experience, and practice is applied with judgment to develop ways to utilize, economically, the materials and forces of nature for the benefit of mankind'. Insofar as social choices or values are considered, they are regarded as important for decisions about the use or development of a technology but not as an integral part of the design process. Social scientists and historians have examined the 'social' character of technology. David Noble (1984, 1977), who contributed some of the first analyses of the social forces shaping engineering designs, challenges the idea of an inner logic driving technology development and leading to specific designs. Tracing the rise of engineering as a profession within the confines of industry (in contrast to other professions which developed as independent practices), he finds that, as a result, the values of engineers reflect those of their employers, only marginally distinct if at all, from managerial objectives. Engineering work, Noble concludes, is oriented toward developing technology that reinforces the existing political and social order.

We have explored until now several ideas and proposals which constitute the formation of technology within the social context. The example with the introduction of matchlocks in Japan, (Noel Perin) showed very clearly the distinction between the social change and the technological push because technological change is responsible for the rapid changes in work and in society. From the time of the first industrial until the third microelectronic revolution, which we are experiencing now, only a few sociological works have attempted to explain the phenomenon of technology in its social construction.

Obvious is also the fact that Marx, (1818-1883), Weber, (1864-1920) and Durkheim ,(1858-1917) who can be seen as the classical sociologists, were all theorists, their findings were based on evidence from historians and not on their own research. Their method is known as the comparative method, that means sociological research involves the comparison of cases or variables which are similar in some respects and dissimilar in others (see also N. Abercrombie et al. 1984). At the same time social surveys were conducted by Charles Booth (1840-1917) using a combination of early survey techniques and other less statistical methods (see McNeill). In the 20th century, the Chicago

School and the anthropologists studied the way of life by living among them and viewing these societies form the inside (participant observation). Following the Second World War, Paul Lazarsfeld (1901-1970) gave greater emphasis on the importance of data, being as objective as possible. Of course in this brief review of methods, it should be said that lots of other methods, ideas and models were taken into consideration testing the ideas in the real world is based as McNeill said:

> The choice of research methods is often decisively affected by choice of topic, and the amount of time, money, and work hours available.

Part Two: Sociological Theory, Economic and Political Environments

From Adorno, (1972) for instance, we gain a focus on 'the application of theory remained uninfluenced by the examining practice. Furthermore theory and empiricism cannot enter the same continuum'. According to the empiricism of technological development processes Hochgerner (1986) argued that it was usual in sociology to take technical equipment and facts into account almost exclusively as societal external factors. Few exceptions existed outside of the dominant development lines of the discipline. The systematic consideration of technical aspects within social facts, the observation of technology as a societal endogenously produced element, implies a transformation of the structures and modes of operation of social relations on a long term basis. Bearing these points in mind, the question of whether the issue, the tasks, the theoretical and methodological points of sociology be extended and partly revived on this foundation? In addition, what is happening with complex positions with which a medium range theory cannot offer satisfactory solutions either. Szèll (1994) focused on the relationship between technology and environment finally arguing that the challenge is to redirect the tools of sociological analysis to the understanding of the special ramification of different social organizations and societies into equations that throw light on both the dual problems of environmental destruction and its control.

Going back to the relevance of sociology of technology, Berka/Hochgerner (1994) who positioned that sociology of technology should not and cannot investigate 'technology' by itself, but rather a technologically structured technical society, and finally that societies structures and features which make technology such a powerful source. A further point has also to be noted. The above ideas, however, could easily suggest that the influence of sociology of technology was significant for the establishment and maintenance of sociology in general. Some may even see these ideas as a reconstruction of the sociological thinking concerning technology. Furthermore, it must be stated here, that, sociology of technology has to establish the limits of sociology; that is to say that it must be forced to move from a peripheral position within the science to a more central place within sociological knowledge. However, the process of defining the 'limits' can not be led from the traditional - the discussion of theories - 'centre' of science. It must be focused on 'research praxis' where exactly these limits are defined and are continually being relocated. Moreover what is also important to note is that if technology can be seen as an element of social action, or as a process in which social relations are to be constituted, then it must enter technology in the theory of social change (Weingart, 1989). Of course the problem of explaining social change is not new, it was a central issue in the nineteenth century sociology. It seems however that the radical social effects of neo-industrialization and technological development of societies are creating fundamental gaps between their existing social systems and the new social evolution of technology. Unanswered is the second part of that time namely the theories of revolution. In a comparative perspective A. Comte, H. Spencer and Emil Durkheim have developed different aspects of evolutionary theory. Theories of revolutionary social change, are particularly deriving from K. Marx who emphasized the importance of class conflict, political struggle and imperialism as the principal mechanism of fundamental structural changes (N. Abercrombie et al., 1994). Following that argument it is obvious that the separation between models of technology as causal factor of social change and models of social determination of technology must be eliminated.

Further, examples of two different perspectives on the social shaping of technology are illustrated, namely, the macrolevel and the microlevel. At the macrolevel, the theory is based on a dynamic view of the social shaping of technology. A proponent of this approach is Salzman (1994), who argued that technology is socially shaped and is part of a larger network of things and people. He and others accepting this view, used this framework, sometimes referred to as a social construction of technology perspective, to build on the traditional studies of science, technology, and society. Thus a number of studies have examined, for instance, the ways in which technology decisions are shaped by nontechnical factors. Research within the emerging field of the social shaping of technology varies quite dramatically in the approaches used, especially in defining the relevant range of social factors. The second perspective, the microlevel, focuses on the usage of technology and is based on the user's perceptions. The conclusion summarizes both the social and the technological impact, and furthermore, it categorizes the research according to varying schools of thought, and shows a new analytical perspective on the sociology of technology on the emerging sociology of information.

Two main factors are key issues for the problems relating to sociology: the individual and the group. Social groups have been defined as collectives of individuals who interact and form social relationships. From the sociology of small group coming to an understanding of larger social collectives, is an effort which has to be shown in sociology, mainly through data. H. Marcuse revitalized and elaborated on Max Weber's perception of the development of the Western world when he labeled modern man as 'one-dimensional'. He argued that the process of rationalization in modern society leads to a more or less systematic elimination of all alternatives. From another point of view, which focuses more on the theory of technological evolution and is based mainly on economic history and anthropology, Bassalla (1988:25) recognizes the larger changes often associated with inventors as well as smaller changes made over a long time period. His theory is rooted in four broad concepts: diversity, continuity, novelty and selection. Diversity can be explained as the result of technological evolution because artifactitious continuity exists, novelty is an integral part of the made world, and a selection process operates to choose artifacts for replication and addition to the existing stock of made things. The argument here is very much related to the first, traditional separation between sociology and technology; i.e., it seems convenient to split up the world into the 'material' and the 'social'.

More precisely, it can be said that theory on the nature and relationship between technology and society is divided into two classic approaches, the technological and the social determinism.

Part Three: Theoretical Considerations and Their Conclusions

Another theoretical approach focussed on political technology and designer technology, which led to actor networks and contingent technology was Elster (1983:9-10) for instance, who underlined two main approaches: first, that technical change may be conceived as a rational goal-directed activity and as the best choice among a set of feasible changes; secondly, technical change may be seen as the cumulative addition of small and largely random modifications of the production process. Freeman (1987) has contributed to this argument, proposing a third approach to technical change. While he recognizes a domain of validity to the other arguments and agrees that they arise from rational choice and changes in production, he argues further that new combinations of radical innovations related both to major advances in science and technology and to organizational innovations could provide a third dimension. He further states that:

such new technological systems can offer such great technical and economic advantages to a wide range of industries and services that their adoption becomes a necessity in any economy exposed to competitive economic, social, political and military pressures. Increasingly in this century, the world-wide diffusion of such new techno-economic paradigms has dominated the process of technical change for several decades and powerfully influences economic and social developments even though it does not uniquely determine them (Freeman 1987:5).

Although the accumulated innovations should actually influence the technical change, no mention is made of existing concepts such as demand pull or technology push and their effect on the technological descriptions. These views, however, can be broken down into two broad categories. First is the theory of the autonomous development of technology (demand pull), posited by those who claim that it is the market and other economic and social influences which primarily determine the scale, rate and direction, and in some cases, even science itself (Freeman 1987:6). Other scholars, such as Schmookler (1966:204) for instance, demonstrated with statistics and figures on patent inventions that the invention activity lagged behind the highs and lows of investment activity. Based on this connection, he wrote that the main stimulus to invention and innovation came from the changing pattern of demand measured by investment in new capital goods. From this point, he went on to argue that external events (proposed to the invention push), for instance, are primarily responsible for the consistency of investments and play the major role in the demand pull theory.

Mowery and Rosenberg pointed out that human needs are almost infinite and often long felt, and cannot explain the emergence of a particular invention at a certain time. They also criticized a series of confusing studies undertaken in the 1960s and 1970s, which attempted to show market demand as the force behind innovation. Eventually, Mowery and Rosenberg came to the conclusion that innovation is the result of the interaction between science and technology push factors. This issue, however, which is based on the autonomous development of technology, has been explained in the past by the author (Katsikides 1994). In this example, the content of this work was to show the interplay between theoretical and philosophical considerations and the autonomy of technological systems. Let me address briefly, the problematic nature of these basic concepts.

Conclusion: The Autonomy of Technology

Reflecting on this problem, Katsikides (1994) on the Autonomous Reflex

System (ARS), had analyzed among other theoretical meanings and techno-social structures. Technology was defined as a system of conditions, which reproduces itself in a quasi-autonomous way (and with increasing compatibility). The system develops itself as an autonomous structure, and reflects symbolism and limited identity within the organizational structures of an enterprise. The ARS dynamics reflect the result of the changes and also its external relations. Reflexive systems proceed more quickly than other sectors or fields of the economy, and thus are able to activate investors, who quickly realize capital returns and profits. ARS affects various activities in the firm because within a short period of time it becomes a tool for rationalization, and as a consequence of which, new compatible structures and standardized norms emerge across all its sectors. Beyond this, ARS demonstrates the dependencies of this system, which lead to coercive detention for investments in new technologies. Investments cause certain rationalizations within ARS, which they cannot be found in machines, but in groupware.

The underlying question, however, is how can the technological development be seen as an independent action while the societal cultural action is regarded as a dependent issue, since they belong to totally different fields of the societal reality? The implication is how and by which methods can theories be shifted or transmitted into societal processes? How does technological development influence organizations? Scientific concepts generally use theories to describe reality, although the quality of these theories always presents a problematic issue. The systematic consideration of technical aspects within social facts, the observation of technology as a societal, endogenously produced element or product, involves the transformation of the structures and modes of operation of social relations on a long term basis. More specific, without the pretension of a complete model, it could argued here that the ARS three-perspective concept which can be applied to various organizational structures, builds a first step in this direction. The first perspective of the model concerns the organizational structure of any enterprise. The second perspective shows the external sphere of action of the enterprise; for example, mother and daughter companies, the relationships to the state, the law, and services. The third perspective covers the internal sphere of action of the organization, for example, trade unions, content of work, working time, collective bargaining, agreements, security, creation of work, corporate identity, enterprises, culture, etc.

For sociology, one possible way out of the methodological dilemma, and the whole approach to it, it could be useful to adopt tools as simulation for instance which by economics is a day in and out business. As Hanappi (1994:171) put it:

It simply means to put models to work. Since models are just mutated copies of those parts of reality, though to be essential for the model-building entity, simulating is testing interaction of these essentials. The most important property of a simulation run is, that it produces interaction results *before* these interactions take place in reality. In other words time is compressed during simulation to enable forecasting of future events. Beyond economies should sociologists also start to simulate social phenomena and in particular technological impact on society is an example par excellence for such research.

Of course, this is not a new idea, Jean Baudrillard pointed out in his books, the *simulation theory,* and showed the radical limits of postmodernism when they are connected with political and social philosophy. His political theory begun within the postmodern philosophy, and positioned that the new world is meaningless and ideologically unchangeable. His central hypothesis was that the social and political elements doesn't constitute anymore the societies' day in and day out business, but on the contrary represents an empty space which reproduces itself without meaning. Simulation is a central key issue by Baudrillard, he underlines that the contemporary human being doesn't experience any reality any more, either passion, nor history, as active subject. Mass media controls life and transforms the individual from the historic perspective into the ecstatic form of information.

With Jean Lyotard and Michel Foucault the main representatives of the postmodern movement one could find similarities in their work. Baudrillards difference is that while by Lyotard and Foucault the presence of the social and the political factors is a prerequisite for their philosophical approach, by Baudrillard their absence creates a methodological start for his simulation theory. Nevertheless, one issue for the build up of simulation societies, is teleworking.

Finally, if we take as an example an organization as a reflection of the social world then we realize quickly that the cultural context creates the vehicle to innovation based on the way we experience reality or the social world. Morgan, (1980) for instance, pointed out the interpretive paradigm very clearly:

Society is understood from the standpoint of the participant in action rather than the observer. The interpretive social theorist attempts to understand the process through which shared multiple realities arise, are sustained, and are chanced.

Notes

1   Whitehead, J. (1994), *How do I improve the Quality of my Management,* in: Management Learning, Vol.25 No.1, p. 149.
2   Abercrombie, N. et al., (1994), *Dictionary of Sociology,* Penguin, p.190, p.76, p. 382, p. 19, p. 12.
3   Holmwood, J. (1995), *Feminism and Epistemology: What kind of successor science?* in Sociology, The Journal of the British Sociological Association, Volume 29, Number 3, Aug. 1995, p. 411.
4   Balnaves M. (1993), *The Sociology of Information,* ANZJS Vol. 29, No.1, March 1993, p.108.

References

Abercrombie N, et al., (1994), *Dictionary of Sociology,* Penguin, London.

Adorno, Th. (1972), *Soziologie und empirische Forschung, in: Adorno et al. Der Positivismusstreit in der deutschen Soziologie,* Darmstadt und Neuwied.

Balnaves M. (1993), *The Sociology of Information,* ANZJS Vol. 29, No.1, March.

Bammè A. et al., (1983), *Maschinen-Menschen, Mensch-Maschinen, Grundrisse einer sozialen Beziehung.* Reinbek.

Basalla G, (1988), *The Evolution of Technology,* Cambridge University Press.

Baudrillard, J. (1978), Die Prazession der Simulakra, in: J. Baudrillard, *Agonie des Realen,* (1978), Marve Verlag, Berlin.

Baudrillard, J. (1983), Der Tod der Moderne, Eine Diskussion, Konkursverlag, Tübingen.

Berka, G., Hochgerner, J., (1994), Social Environment of Technical Progress, in: Katsikides et al., Patterns of Social and Technological Change in Europe, Avebury, Aldershot.

Bijker, W, et al., (1987), *The Social Construction of Technological Systems,* Cambridge/Mass.-London.

Elias, Norbert, (1978), *Was ist Soziologie,* München, p.142.

Ellul Jacques, (1964), *The Technological Society,* Vintage Books, New York.

Elster Jon, (1983), *Explaining Technical Change,* Cambridge University Press, Cambridge.

Freeman Chr. (1987), *The Case of Technological Determinism,* in: R. Finnegan, et al. *Information Technology: Social Issues,* The Open University, Hodder&Stoughton, pp. 5-6.

Habermas, J, (1972), *Technik und Wissenschaft als Ideologie,* Frankfurt am

Main.

Hanappi, H. (1994), *Evolutionary Economics*, Avebury, Aldershot, p. 171.

Hochgerner, J, /Berka, G, (1994), *Social environment of technical progress*, in Katsikides et al., *Patterns of Social and Technological Change in Europe*, Avebury, Aldershot.

Hochgerner, Josef. (1986), *Arbeit und Technik. Einführung in die Technksoziologie*, Kohlhammer Verlag, Stuttgart.

Hülsmann Heinz, (1985), *Die technologische Formation*, Berlin.

Gehlen, A, (1986), *Die Seele im technischen Zeitalter. Sozialpsychologische Probleme in der industriellen Gesellschaft*, Reinbek.

Katsikides, S. (1997). *Sociology and the Functions of Technological Autonomy*, in Innovation: The European Journal of Social Sciences, Volume 10, Number 2, pp. 195-201.

Katsikides, S, et al., (1994), *Patterns of Social and Technological Change in Europe*, Avebury, Aldershot.

Katsikides, S. (1994), *Informatics, Organization and Society, Oldenbourg*, Wien-München.

Lange, W.D. (1975), *Französische Literaturkritik der Gegenwart*. Kröner, p. 248.

Linde, H, (1972), *Sachdominanz in Sozialstrukturen*, Tübingen.

Linde, H, (1982), *Soziale Implikationen technischer Geräte, ihrer Entstehung und Verwendung*, in R. Jokisch (Hg.) *Techniksoziologie*, Frankfurt am Main.

Loewith, K, (1964), *Das Verhängnis des Fortschritts*, in: H. Kuhn, F.Wiedmann (Hg.) *Die Philosophie und die Frage nach dem Fortschritt*. Verhandlungen des Siebten Deutschen Kongresses fuer Soziologie, München.

Marcuse, H. (1967), *Der eindimensionale Mensch*, Neuwied-Berlin.

Marcuse, H.(1941), *Some Social Implications of Modern Technology*, in Zeitschrift für Sozialforschung, Jg.9, Nr.3/41.

McNeill M. (1989), *Research Methods*, Routledge, 2. edition.

Morgan, G. (1980), *Paradigams, Metaphors and Puzzle Solving in Organization Theory'*. Administrative Science Quarterly 25:608-609.

Mowery, D. and Rosenberg N, (1979), *The influence of market demand upon innovation*, Research Policy, Vol.8 No.2.

Mumford, Lewis. (1977), *Mythos der Maschine*. Fischer-alternativ-Verlag, Frankfurt a.m. p. 493.

Noble David, (1977), *America by Design*. Oxford University Press.

Ogburn, W, (1969), *Kultur und sozialer Wandel*, Neuwied/Berlin.

Rammert. W, (1983), *Soziale Dynamik der technischen Entwicklung*, Obladen.

Rammert. W, (1991), *Entstehung und Entwicklung der Technik, Stand der*

Forschung zur Technikgenese in Deutschland, WZB, Berlin.

Rammert. W, (1993), *Technik aus soziologischer Perspektive,* Obladen.

Salzman H, Rosenthal S, (1994), *Software by Design,* Oxford University Press, New York.

Salzman, H. (1994), *The Social Context of Software Design,* in: Katsikides, S. (eds.) (1994), *Informatics, Organization and Society,* Oldenbourg, Wien-München.

Salzmann H, Rosenthal, S., (1994), *Software by Design,* Oxford University Press, New York.

Schmookler, J, (1966), *Inventions and Economic Growth,,* Mass. Harvard University Press, Cambridge.

Széll G. (1994). *Sociology: State of the Art II, Political, economic and socio-demographic dimensions,* in: International Social Science Journal, Blackwell Publishers, Vol. 140, xlvi, No. 2, June 1994.

Széll G., (1994), *Technology, production, consumption and the environment,* International Social Science Journal, June 1994 Vol. 140, Blackwell, Oxford.

Tschiedel, R, (1989), *Sozialverträgliche Technikentwicklung,* Obladen.

Tschiedel, R, (1990), *Die technische Konstruktion der gesellschaftlichen Wirklichkeit,* München.

Ulrich, O, (1979), *Technik und Herrschaft,* Frankfurt am Main.

Weingart, P, (1989), *Technik als sozialer Prozess,* Frankfurt am Main.

Willkes H. (1989), *Systemtheorie entwickelter Gesellschaften,* Juventa, München.